DESTINATION MOON

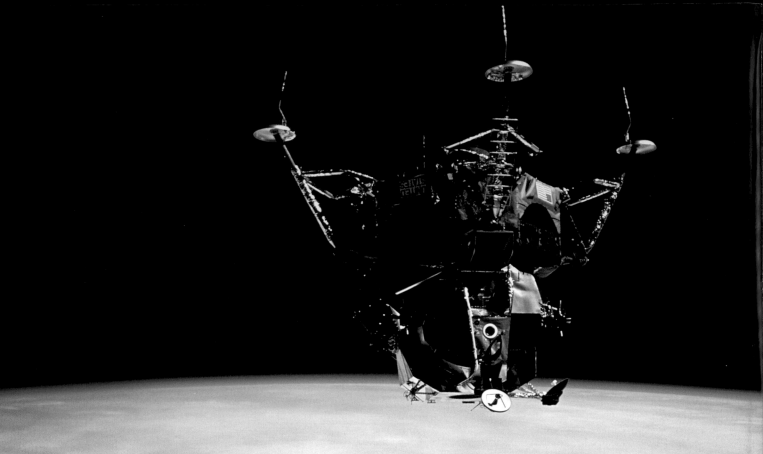

THIS IS A CARLTON BOOK

Design copyright © 2005 Carlton Publishing Group
Original text and text compilation copyright © First
Person Productions LLC 2005
NASA astronauts' mission
dialogue and interviews copyright
© NASA 2005

This edition published in 2005 by Carlton Books Ltd
A Division of the Carlton Publishing Group
20 Mortimer Street
London
W1T 3JW

A CIP catalogue for this book is available from the
British Library.

ISBN 1 84442 712 9

Project Editor: Amie McKee
Art Editor: Vicky Holmes
Design: Simon Mercer
Picture research: Rod Pyle
Production: Lisa Moore

Printed and bound in Dubai

The Apollo Missions in the Astronauts' Own Words

DESTINATION MOON

Rod Pyle

FIRST PERSON PRODUCTIONS   In association with

CARLTON BOOKS

CONTENTS

INTRODUCTION

Looking back across the last century, one seeks a bright spot in the dark timeline of war, strife and conflict. Then in the mid 1960s, between 1961 and 1972, the Apollo Program shines forth as mankind's greatest peacetime achievement. Focused, intense, and powerful, it was, perhaps, America's finest hour of the twentieth century.

For those of us old enough to remember, the Apollo years were a wondrous time. Deep in the gathering darkness of the Vietnam War, the United States and the Soviet Union played out a proxy war in the skies overhead. Rather than the more direct, and deadly, land wars these two countries fought on Third-World soil, the race to the moon was essentially a peaceful endeavor accelerated by the imperative of national prestige and the natural competition between capitalism and communism. In this, as in so many other areas, democracy won. It was a heady time.

The forces that propelled Apollo moonward were vast and varied. When John Kennedy made his announcement in 1961 that an American would land on the moon, the US had just flown Alan Shepard in a 15-minute, suborbital toss aboard an underpowered Redstone missile. Just prior to that, the USSR had sent Yuri Gagarin into space to orbit the Earth. The Russian rockets were large (in comparison) and powerful, while the Redstone was less powerful than the Apollo Command Module escape rocket. It was not a promising start.

By 1967, the stakes were high and the forces were better matched. The Soviets were developing a huge booster and a lunar lander (for one man), and the US was working overtime on Apollo. The US lost three crewmen to an accident on the ground, and soon the Soviets would see their magnificent, if flawed, N1 moon rocket explode on the pad, killing a number of top engineers and designers. It was the end of their manned moon effort.

From 1968 onwards, the trip to the moon was a uniquely American affair. Twenty thousand contractors and over 400,000 people would ultimately conspire to accomplish Kennedy's mandate just before the end of the decade. It was an unlikely alliance of scientists, engineers, corporate giants and government. Perhaps the most profound innovations came in the largely unheralded discipline of management.

Rarely had a challenge of such scale been attempted, and certainly not in peacetime. The only rational comparison might be with the Manhattan Project, but ultimately even that great enterprise pales in comparison. The men who managed the Apollo Program invented a whole new way of doing business. Inspired by the meticulous methodology of Wernher von Braun's German rocket team that was transplanted to the US after World War II, Chris Kraft, Sam Phillips, Jim Webb, Tom Paine, George Low, Gene Kranz and many others virtually revolutionized how great endeavors were accomplished. Without this, there would have been no manned moon landings.

When the citizens of Earth gathered around their mostly black-and-white television sets on July 20, 1969, they were united in wonder for two-and-a-half priceless hours. Neil Armstrong and Edwin Aldrin changed how they thought about their world, and beyond, during that short moonwalk.

Subsequent missions continued to astound the faithful. Although the television networks abandoned Apollo for more lucrative reruns (with the exception of the newsworthy Apollo 13), there were those of us who could still just not get enough. We read the papers and listened to the radio. Later, we read the books and watched the documentaries. The fine works of Al Reinert (*For All Mankind*) and Blaine

Bagget (*Spaceflight*) come to mind. Still later we enjoyed Andy Chaikin's fine book, *A Man on the Moon*, and the Tom Hanks miniseries distilled from that tome.

There is a vast fascination with space exploration today. The Jet Propulsion Laboratory's interplanetary missions continue to captivate, and the Space Station orbits overhead every 90 minutes, promising great interplanetary voyages in the near future. Even the aging Space Shuttle is still an inspiration. But none of these is quite the measure of the intensity and excitement of the Apollo years.

So it is in that spirit *Destination Moon* is presented. In these pages are contained many of the finest photographs shot during all the missions to the moon, culled from thousands of original negatives. The astronauts are presented, for the first time, in their own words with insightful editorial commentary. Interviews with the astronauts and project leaders complete this new history of the Moon programme. The overall effect is to transport the reader back to that time of breathless wonder to enjoy (or re-experience) the Apollo Program as we who were there did.

It was, truly, the best of times.

Rod Pyle, Pasadena, California 2005

10 THE FIRST VOYAGE

Preparing to Go

When President John F. Kennedy announced in 1961 that America would go to the Moon, the National Aeronautics and Space Administration (NASA) had to find a way to get there. The path was winding and tortuous, but in eight short years the task was done. One tragedy befell them, but NASA's ingenious engineers and managers pulled the space program out of the ashes of Apollo 1 and moved forward.

The Sea of Tranquility is misnamed. Despite its poetic title, it is a vast, bleak plain of gray rocks and dust, punctuated by craters, boulders and rocky fragments. Above it, the Sun shines like a 40-million-watt spotlight, bathing the entire region in a 250° Fahrenheit (121° C) blast-furnace glare. But this is, after all, the Moon.

On the southern edge of this spare expanse, a visitor would encounter a strange sight. Sitting alone in a flat field, next to a few small craters and not far from a field of huge rocks, is a gold foil-wrapped contraption that looks like an uncovered, high-tech bandstand. Supported by four spindly, tubular legs, the artefact has not aged noticeably since 1969.

America's target: the Moon.

There is a barely visible scorching of the rocks beneath it. Nearby, a red-white-and-blue flag wired to a pole lies incongruously in the faded dust. This is a future museum site; a monument to humankind's greatest act of daring. It is Tranquility Base – the place where humankind first landed on another world, during an adventure called Apollo.

Eight years before this odd-looking relic broke the Moon's five-billion-year silence, the Apollo Program was an unknown. Some had thought about interplanetary travel, but the notion was far from reality. It was not until a muggy May afternoon in 1961 that the dream took a definite shape, as President John F. Kennedy addressed a joint session of the United States Congress.

John F. Kennedy addressed a joint session of Congress on May 25, 1961. His announcement of a daring lunar program was regarded variously as spectacular, reckless or just plain crazy. Chris Kraft, Flight Director at NASA, later said he thought that Kennedy had "lost his mind."

"I believe that this nation should commit itself to achieving the goal, before this decade is out, of landing a man on the Moon and returning him safely to the Earth. No single space project in this period will be more impressive to mankind, or more important for the long-range exploration of space; and none will be so difficult or expensive to accomplish...."

The speech continued, expressing concern about vast political complications and national needs. However, its impact had already been felt. Humanity, in the guise of the United States, was reaching out into space and would travel to another world.

Apollo 1

Kennedy's proclamation had set America on the fast track to make the dream of extended spaceflight a reality. The Soviet Union had beaten the United States to achieve several firsts in the space race. They launched the first successful satellite into Earth orbit (*Sputnik 1* on October 4, 1957); sent Yuri Gagarin into orbit in Vostok 1 in April 1961. And, in 1965, a Russian cosmonaut, Alexei Leonov, became the first man to spacewalk. It seemed that they might also be the first to reach the Moon.

Space was suddenly the highest of the high ground. On a scale unseen since the Manhattan Project – the Second World War program aimed at developing nuclear weapons – men of science and industry,

academics and engineers, politicians and dreamers, in short, people from all walks of life, united in an unlikely alliance to reach to the Moon. It was a time like no other.

That expression of raw willpower led to the development of the first Apollo spacecraft, known informally as Apollo 1, which was first set on top of a rocket in late 1966. On a fateful day in January 1967, three astronauts sat within the metal craft. The rocket was not fueled and the usual streaming vapors did not trail from its innards. There were no crowds surrounding Cape Kennedy in Florida. It was just another test, another small step towards the "Big Goal". Nothing seemed unusual. And although more than one astronaut had complained that the spacecraft was substandard and potentially dangerous, this was not on anybody's mind at the time. The crew was just trying to make the radio work.

Gus Grissom, who was slated to be the commander of the first Apollo mission into Earth orbit, sat beside his Lunar Module (LM) pilot Ed White and his Command Module (CM) pilot Roger Chaffee. Gus was not a happy man. "How are we going to get to the Moon," he snapped, "if we can't talk between three buildings?" The controllers in the nearby launch control had heard it before. Gus could be a curmudgeon.

Something they had never heard before came shortly afterward. "We've got a fire in the cockpit!" Thirty seconds later the men had been scorched and asphyxiated. Within a minute, the hull of the capsule had been torn open by extreme heat and a build-up of pressure. Smoky flames erupted. But still, the launch pad

The prime crew for Apollo 1: from left, Virgil "Gus" Grissom, Roger Chaffee and Ed White. Note the early, and flammable, Apollo spacesuits.

The horrific results of the Apollo 1 inferno. There were so many flammable plastics in the spacecraft that the bodies of the crew had to be removed a little at a time.

The funeral procession for the crew of Apollo 1. John Glenn, who was one of the original seven astronauts selected to participate in the space program, can be seen in the center of the image. Behind him are Gordon Cooper and John Young.

technicians were unable to unscrew the many bolts that held the hatch in place like the door to a tomb. Within minutes, Grissom, White and Chaffee were dead.

Even before the men were buried, recriminations flew within NASA and the industry that supported it. But the fault was really collective. America was trying, with primitive Cold War technology, to build the first spaceship to fly from Earth to another celestial body. It was a vast undertaking, and one fraught with risks. Chance had caught up with – and surpassed – luck. Now those involved had to ensure such a tragedy would not be allowed to happen again.

As it turned out, the fire was a demon waiting to snatch the souls of the three men. The capsule, as was standard practice, had been pumped full of pure oxygen for the test. Inside it were loose connections, frayed and wrongly routed wiring and lots of flammable materials such as nylon and Velcro. Under normal circumstances, these matters were, at the very least, annoyances and, at the most, issues that needed to be resolved. In a pressurized, pure oxygen environment they comprised a bomb. When a spark crackled from one of the wire bundles beneath the astronauts' couches, their time was up. Nothing could have stopped that fire. It was as if a hand had reached up from hell and crushed the living in Apollo 1.

Over the next 18 months, the space program was assailed from both inside and out. Investigations were undertaken, and some politicians tore into NASA, seeing this as an opportunity to end the expensive Apollo missions. But equally powerful personalities were operating on the other side of the argument, notably the astronaut Frank Borman, who had commanded the Gemini VII mission in 1965 and was later to become the commander of Apollo 8. His force of will, as much as anything, contributed to getting Apollo back on track. Few wanted to say no to an astronaut.

As work accelerated on the improved Command Module, there were still uncertainties. How do you perfectly protect over 15 miles (24 kilometers) of wire inside one small craft? And then there was the Saturn V rocket, which was to boost the Apollo crews into space. At 363 feet (111 meters) this rocket was taller than the Statue of Liberty and 13 times heavier. It was filled with a million gallons of highly flammable fuel, and if it malfunctioned on the pad, it would explode with the force of a small nuclear bomb.

Apollo: Testing in Orbit

On October 11, 1968, Apollo was deemed ready to launch. And at this point in the 1960s, if NASA was to make Kennedy's deadline of "before this decade is out," Apollo had to fly.

On a hot October morning in 1968, Apollo 7 launched into a blue Florida sky. The Saturn V was still undergoing tests, so this mission used the smaller, older Saturn 1B rocket. After 163 orbits in 10 days the crew prepared to re-enter the Earth's atmosphere and return home.

The crew of Apollo 7 landed safely back on Earth and the flight was a success. The Command Module and larger Service Module containing the rocket unit

The Apollo 8 Command-Service Module (CSM) structure, with the Block II Command Module at the top.

Crew: Apollo 7

The commander of Apollo 7 was Walter M. "Wally" Schirra. Born in New Jersey in 1923, he graduated from the US Naval Academy in 1945 and Naval Flight Training in 1947. He joined NASA in 1959 as one of the original seven astronauts recruited for the Mercury Program, which aimed to send the first American astronauts into space and return them to Earth. After flying on Mercury 8 in 1962, Schirra went on to work on the Gemini program. Now he was to command the first piloted Apollo flight.

Walter "Walt" Cunningham occupied Apollo 7's second seat. He was born in 1932 in Iowa, and received a Bachelor of Science degree with honors in Physics in 1960 and a Master of Science degree in Physics in 1961 from the University of California at Los Angeles. He joined the US Navy in 1951, moving to a Marine squadron in 1953. Cunningham served on active duty with the Corps until August 1956. He signed on with NASA in 1963.

Donn Eisele, Apollo 7's Command Module pilot, was born in Ohio in 1930. He received a Bachelor of Science degree from the United States Naval Academy in 1952 and a Master of Science degree in Astronautics in 1960 from the Air Force Institute of Technology. He accrued thousands of hours in jets after a stint at the test pilot school, before joining NASA in 1963. On Apollo 7 he was to be the first to pilot the most complex flying machine ever devised.

The crew of Apollo 7. From the left: Walt Cunningham, Donn Eisele and Wally Schirra. Note the redesigned and flameproof Apollo suits.

behind the capsule had all checked out. The crew had fired the rocket eight times, to ensure that it would work in orbit around the Moon.

With all systems go, there was a new debate within NASA. Long ago they had agreed to a stepped development program: test one unit, and if it passed, test the next. The German scientists at NASA, among whom was Wernher von Braun, a key player in the Apollo Program, did not like to fly untried spacecraft in critical roles. Nevertheless, powerful voices within the space agency wanted to fly an "all-up" mission – that is, to go right for the Moon. For while the goal of landing on the Moon before 1970 was in sight, the Soviet Union was known to be working on a lunar landing program of their own, which meant that time was of the essence. Even if the Russians couldn't land there, they might attempt to orbit the Moon, just to steal America's thunder once again. NASA was not about to let that happen.

Into the Void: Apollo 8

So on December 21, 1968, Apollo 8 roared into history on top of the new Saturn V rocket. It was only the second flight of the new rocket, and its most recent, unpiloted, mission had experienced major problems. But President Kennedy's clock was ticking, so up they went. Onboard Apollo 8 were three very different men; it was one of NASA's most diverse crews.

As they departed Earth orbit to head for the Moon, the Apollo 8 crew would forego one key element of all later Apollo flights. In the cavernous stowage bay behind the capsule was nothing but empty space. In later flights, it was where the Lunar Module (LM) would go. But in December 1968, the LM was not ready for service, so the mission was to orbit the Moon. The only real problem anyone could think of would be if, for some reason as yet unforeseen, the combined Command and Service Modules (CSM) failed. But that was considered inconceivable, at least until NASA learned differently from Apollo 13 (see p.85).

But the gods smiled on Apollo 8. After the three men became the first to escape the Earth's gravitational pull, they began the long fall into the Moon's. As Jim Lovell maneuvered inside the cramped cabin of the CM, he had his first real problem of the mission.

From below the astronauts' couches there came a popping noise, followed by a low hiss.

01:13:38 **Borman:** "What was that?"
01:13:40 **Lovell:** "Uh... my life jacket."
01:13:41 **Borman:** "No kidding?"
01:13:45 **Lovell:** "It hooked on the tank here. It flicked up."
01:13:52 **Borman:** "Is it blowing up?"

Lovell is too embarrassed to answer right away.

01:13:53 **Lovell:** "It's too early [for splashdown]."
01:13:57 **Anders:** "Why don't you take it off and give it to me, and I'll try to take it apart while you watch the panel."
01:14:05 **Anders:** [To Houston] "Lovell just caught his life vest on Frank's strut..."

Lovell had managed to inflate his life vest, to the extreme amusement of his crew mates. The capsule was already small; with the vest inflated he felt like a sausage. As they cruised into the vast night of space, the Moon loomed ever larger in their windows.

Two days later they were preparing to make their Lunar Orbit Insertion (LOI) burn. This would happen when they were behind the Moon, when they would be the first humans to see the far side. It was the most helpless moment at Mission Control, for they were not able to talk to or hear from the astronauts during this period of blackout. As Apollo 8 sped into the area behind the Moon, where they would lose their signal, Mission Control at Houston said their temporary goodbyes.

68:57:06 **CapCom:** "Apollo 8, Houston. One minute to LOS. All systems go... safe journey, guys."
68:57:24 **Anders:** "Thanks a lot, troops. We'll see you on the other side."

Shortly thereafter, Apollo 8 passed behind the Moon, setting yet another new record. It was the first flight to be completely out of communication with the Earth.

Christmas Eve 1968: In Orbit Above the Moon

On the fourth day of the flight, in the greatest PR coup of NASA's short history, each of the crew read from the Bible's Book of Genesis. Their words traveled to a world 243,000 miles (391,000 kilometers) distant whose population was transfixed, whatever their religion. Bill Anders started the event:

71:40:00 **Anders:** "For all the people on Earth the crew of Apollo 8 has a message we would like to send you. 'In the beginning God created the Heaven and the Earth.'"

Crew: Apollo 8

Frank Borman was Apollo 8's mission commander. Born in Indiana in 1928, Borman received a Bachelor of Science degree from the US Military Academy at West Point, New York, in 1950, followed by a Master of Science degree from the California Institute of Technology (Caltech) in 1957. He joined NASA in 1962 after serving as a test pilot and instructor. He was as no-nonsense an astronaut as ever lived.

Jim Lovell was also born in 1928 in Ohio. He joined NASA in 1962 with a Bachelor of Science degree from the US Naval Academy. He had also graduated from the Naval Test Pilot School and flown thousands of hours in jets. Lovell and Borman were both spaceflight veterans, having flown together in Gemini VII in 1965.

Bill Anders was born in Hong Kong in 1933. He received a Bachelor of Science degree from the US Naval Academy in 1955 and a Master of Science degree in nuclear engineering from the Air Force Institute of Technology. He flew with the Air Force in an all-weather attack squadron until being selected by NASA in 1964. Anders replaced Mike Collins, who had fallen ill shortly before the mission.

A prime crew in front of NASA's Vehicle Assembly Building (VAB).
From the left: Jim Lovell, Bill Anders and Frank Borman.

Anders continued until Lovell took over to read his part. Finally, Frank Borman took his turn:

"And God said, 'Let the waters under the heavens be gathered together unto one place, and let the dry land appear.' And it was so. And God called the dry land Earth; and the gathering together of the waters called he Seas: and God saw that it was good... And from the crew of Apollo 8, we close with good night, good luck, a Merry Christmas and God bless all of you – all of you on the good Earth."

After 10 orbits around the Moon the crew of Apollo 8 fired their engines and broke free of lunar gravity to head back home. Some time later, the Earth's gravitational field took hold and they sped toward splashdown. They arrived safely on Earth 147 hours after launch.

America had flexed its technological muscles. They had reached the Moon and orbited it. NASA could have stopped there – and few would have blamed them. But it was not enough. They wanted the big prize – a Moon landing.

The curse of Apollo 1 was broken and the three martyred astronauts – Grissom, Chaffee and White – traded their lives for the successes of the later Apollo flights. With the success of Apollo 7 and Apollo 8, the Moon seemed very close indeed.

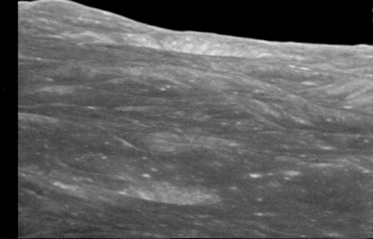

"The good Earth."

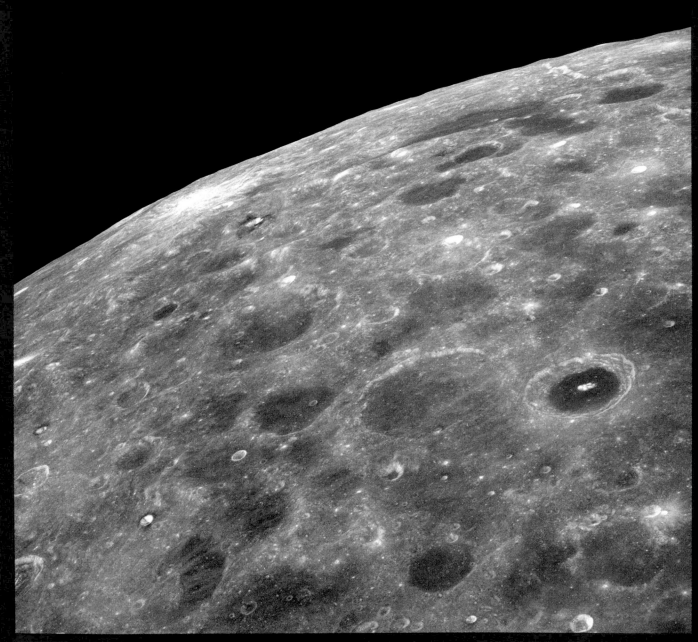

Orbiting the Moon in Apollo 8.

Retrieving the Apollo 8 capsule after splashdown.

09 CLOSE ENOUGH TO TOUCH

March 1969: In Earth Orbit

After the triumphant flight of Apollo 8, there remained two crucial tests before landing on the Moon. First, Apollo 9 was to test the new spacesuits and the Lunar Module in Earth orbit. This was to be followed by Apollo 10, which would take the LM to the Moon and run it through its paces in lunar orbit.

Spider and Gumdrop

Considering that this was the first flight test of a LM, the mission had an irreverent tone to it. In the grand tradition of naming spacecraft, the crew of Apollo 9, James McDivitt, David Scott and Russell "Rusty" Schweickart, decided to ignore romance and go straight for literal description. The CM would be known as *Gumdrop* and the LM, *Spider*.

Apollo 9 was, after Apollo 13 (see p.84), arguably the toughest mission of the entire program. Jim McDivitt might have agreed, but he would not have

On a March morning in 1969, Apollo 9 soars into the sky over Florida. This was only the second piloted flight of the Saturn V rocket.

told you so. Ever the consummate pilot and scientist, McDivitt's goal was to master the LM.

The Saturn V was also still a test piece in terms of human transport, with only the Apollo 8 mission behind it. The combined Command and Service Module, with the capsule at the top and the cylindrical rocket engine–life support combination beneath it, had flown twice previously. But the LM, just now starting to trickle out of production at Grumman Aerospace in New York, was another story.

This unlikely-looking spaceship, the product of enough out-of-the-box engineering to fill the cavernous Vehicle Assembly Building at Cape Canaveral, was a new generation of flying machine. Tremendously lightweight, and fragile as well, it was designed to fly only in space.

Crew: Apollo 9

Born in Chicago in 1929, James McDivitt received a Bachelor of Science Degree in Aeronautical Engineering from the University of Michigan in 1959. He joined the Air Force in 1951 and attended the US Air Force Experimental Test Pilot School after flying more than 140 combat missions in Korea. McDivitt also flew as an experimental test pilot at Edwards Air Force Base in California before joining NASA in 1962. His first spaceflight was on Gemini 4 in June 1965.

With McDivitt on this groundbreaking voyage was Rusty Schweickart, the red-headed astronaut who joined NASA in 1963. He had a Bachelor of Science degree in Aeronautical Engineering and a Master of Science degree in Aeronautics and Astronautics from Massachusetts Institute of Technology (MIT).

The Command Module Pilot was David Scott, another Gemini veteran. He joined NASA in 1966, after receiving a Bachelor of Science degree from the United States Military Academy and the degrees of Master of Science in Aeronautics and Astronautics and Engineer in Aeronautics and Astronautics from MIT.

Apollo 9's crew. From the left: Jim McDivitt, Dave Scott and Rusty Schweickart.

Spider in Earth orbit on Apollo 9. This was the first flight of a Lunar Module and it performed brilliantly.

The craft was not aerodynamic, and it could not withstand the rigors of operation in Earth's atmosphere. It was designed purely to land on and take off from the Moon.

The LM's lightly built structure had some astronauts calling it the "tissue paper spacecraft"; indeed, when pumped-up to operational pressure, its sides bulged. When a technician working inside the LM had dropped a screwdriver a few months previously, it had punctured and damaged the hull. And the hatch... well, you could peel it open if you weren't careful.

These and other thoughts filled the minds of the crew of Apollo 9, who were charged with orbiting a Command Service Module and a Lunar Module, practicing docking and undocking the craft and flying free. The mission was also going to be the first test of the new Apollo spacesuit in a hard vacuum.

The three astronauts who made up the crew of Apollo 9 would "hang it out over the edge" on this voyage. And Jim McDivitt would not have had it any other way.

The mission launched on March 3, 1969. The crew orbited the Earth with little drama, and separated the CSM from the S4B stage, then turned 180 degrees to retrieve the LM from its metal cocoon. After docking with the LM, McDivitt and Schweickart moved to the lunar lander, ran through numerous checklist items, then undocked from the CSM and put the new machine through its paces. It was the first time a crew had flown a craft without a heat shield around the Earth. They flew up to 111 miles (178.6 kilometers) from the CM, and then proceeded to rendezvous and dock with it. Later, they opened the front LM hatch to the hard vacuum of space and put the Apollo lunar suits through their paces by climbing down the LM ladder into the great void of orbit. After 10 days and 151 orbits, the flight ended. It was a great success. The lunar hardware that had been tested exceeded expectations. Now NASA needed just one more simulation before landing on the Moon. This time it was to be in lunar orbit.

Barnstorming: Apollo 10

A few months later it was the turn of Eugene Cernan and Thomas P. Stafford. The CSM had been checked out in lunar orbit on Apollo 8, but the LM still needed testing in the vicinity of the Moon. There was also the issue of the so-called mascons, or mass concentrations, which were areas surrounding the Moon that had a higher gravitational pull. NASA wanted one more flight to check out this rather lumpy gravitational field before committing to a landing.

Apollo 10 launched on May 18, 1969, just three months before the flight of Apollo 11. After the rocket booster's second stage dropped free, the S4B stage ignited and sent the crew into Earth orbit. However, when they fired the S4B to leave Earth orbit, the rocket developed a severe vibration. It became so violent that the three astronauts could barely read their instruments, and Stafford came close to aborting the flight. Luckily, the S4B shut down at the appointed time, the vibrations ceased and Apollo 10 was on its way to the Moon.

Upon arriving in lunar orbit, Cernan and Stafford moved down the tunnel connecting the CM and LM. After powering up the LM they prepared to undock

Rusty Schweickart stands on the LM's "front porch" in Earth orbit. It was the first test of the Apollo spacesuit in a hard vacuum.

Crew: Apollo 10

Commanding Apollo 10 was Thomas P. Stafford, a veteran of the space program, having flown on both Gemini VI in December 1965 and Gemini IX in June 1966. Born in Oklahoma in 1930, Stafford received a Bachelor of Science degree from the US Naval Academy in 1952, and was then granted his pilot wings at Connally Airforce Base, Waco, Texas, in September 1953. He joined NASA in 1952, after four years as a test pilot.

Stafford's Lunar Module pilot was Eugene "Gene" Cernan, who later commanded Apollo 17 (see pp.156). Born in Chicago in 1934, he received a Bachelor of Science degree in Electrical Engineering from Purdue University, Indiana, in 1956 and a Master of Science degree in Aeronautical Engineering from the US Naval Postgraduate School, Monterey, California. Cernan flew more than 5,000 hours in jet aircraft before being selected by NASA in 1963.

John Young was the Command Module Pilot. Born in 1930 in San Francisco, he earned a Bachelor of Science degree in aeronautical engineering with highest honors from Georgia Institute of Technology in 1952. After test pilot training at the US Navy Test Pilot School in 1959, he was assigned to the Naval Air Test Center for three years. Young was recruited by NASA in 1962 and later commanded Apollo 16.

The crew of Apollo 10. From the left: Gene Cernan, Tom Stafford and John Young.

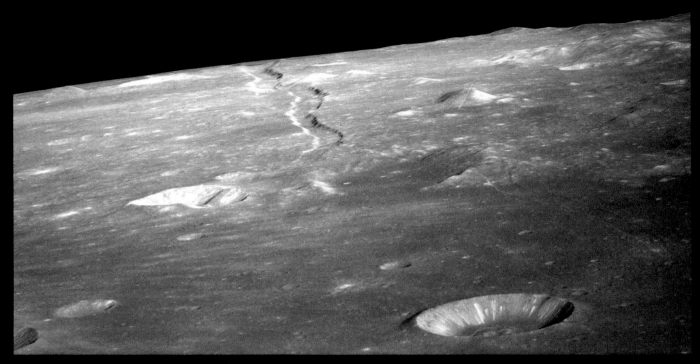

Skimming the Moon in Apollo 10. This is an oblique view of Rima Ariadaeus.

and barnstorm the Moon. They fired the LM's descent engine and dropped from about 70 miles (113 kilometers) to 5,000 feet (1,500 meters) above the lunar surface. The LM then skimmed around the Moon at a height of about nine miles (14.5 kilometers), its crew awestruck by the scenery. Some of the mountains that they flew past were only a few miles beneath them.

This phase of their mission completed, Cernan and Stafford now had to jettison the descent stage and use the lunar ascent engine to return to the Command Module. The guidance computer flashed the code asking if they wanted to proceed. Just as Stafford hit the button and dropped the descent stage, the LM spiraled out of control.

It took almost 10 seconds to regain control of the craft. Cernan later calculated that they were about two seconds away from smashing into the side of one of the lunar mountains. After regaining their composure, Cernan and Stafford fired the ascent engine and flew back up to rendezvous with John Young in the CM. Then they headed for home.

These were the pathfinding flights. Apollo 9 was the full-up test for the Apollo spacesuits and the Lunar Module. Apollo 10 was the final rehearsal for navigating in lunar orbit and docking, undocking, LM staging and returning to orbit. Without these missions, Apollo 11 might not have been the staggering success reflected in the history books. And now... it was time to go for broke.

Opposite: Recovery: All returning Apollo crews had to be prepared to ride out the oceans for some time before being picked up by the US Navy.

08 APOLLO 11 – FOOTPRINTS

July 20, 1969: The Moon

The Apollo 11 voyage fulfilled the dream of putting a man on the Moon. However, the Moon did not reveal her secrets with ease. This momentous mission was nearly aborted at least once, and it was only through sheer determination and bravery by those involved both on the ground and in space that the mission succeeded.

High Above *Mare Tranquillitatis*

The vast, sun-blasted plains of the Sea of Tranquility, or *Mare Tranquillitatis* to give it its Latin name, stretched as far as the eye could see. Hundreds of miles wide, it seemed a perfect target for the first lunar landing by an Apollo crew. It was smooth, flat and relatively unmarred by dangerous craters.

Above it, a speck of white, accented by flashes of gold, appeared on the horizon. Slowly it grew, revealing an oddly shaped spacecraft, looking for anything like a robotic spider. This was only the second time such a

The Sea of Tranquility as photographed from orbit. Although it looked like a smooth, safe place for the first landing, there were dangerous boulder fields and craters that only became apparent when viewed from a lower altitude.

craft had orbited this airless body in its five-billion-year history. This one would attempt a landing. Below the fast-moving craft, the lunar surface remained cold and silent.

Not so in the skies above, where the Lunar Module *Eagle* was transitioning from a calm orbit to a very eventful landing. Inside were Neil Armstrong and Buzz Aldrin, two of the best astronauts NASA had to offer. It would take every bit of their skill and training to complete this, the first Apollo landing on the Moon. In the cramped cabin, an alarm rang shrilly in the astronauts' headsets.

102:38:26 **Armstrong:** "Program Alarm!"

At Mission Control in Houston their CapCom, Charlie Duke, himself an astronaut, attempted to reassure them.

Crew: Apollo 11

Neil Armstrong was the soft-spoken commander of Apollo 11. Born in 1930 in Ohio, he attended Purdue University, Indiana, earning his Bachelor of Science degree in Aeronautical Engineering in 1955. During the Korean War, he flew 78 combat missions in jet fighters, and was awarded the Air Medal and two Gold Stars. He later earned a Master of Science degree in Aerospace Engineering from the University of Southern California. After a stint at Edwards Air Force Base as a test pilot, he joined NASA in 1962, part of the second class of astronauts. At this point, he was already considered an astronaut by the Air Force, for flying the X-15 jet above the atmosphere at the "edge of space."

Edwin "Buzz" Aldrin was also born in 1930, in New Jersey. He received a Bachelor of Science degree in 1951 from the United States Military Academy at West Point, New York. He later earned a Doctorate of Science in Astronautics from MIT, completing his dissertation on "Guidance for Manned Orbital Rendezvous." This would come in handy both for mission planning and during his mission on board Gemini XII in November 1966. Aldrin was a very focused man, whose inward nature was an interesting complement to Armstrong's shyness. Together, they made up the most intensely watched team of explorers in history.

Their colleague, who orbited overhead in the CM, was Michael Collins, born in Rome, Italy, in 1930. He also graduated from West Point with a Bachelor of Science degree, then went on to become a test pilot at Edwards Air Force Base in California. He joined NASA in 1963. More outgoing than Armstrong, he was very much the philosopher of the voyage.

NASA's Apollo 11 flight crew, Neil A. Armstrong (Commander), Mike Collins (Command Module pilot) and Buzz Aldrin (Lunar Module pilot) stand near the Apollo—Saturn V space vehicle that carried them into space on July 16,1969.

102:38:28 **CapCom:** "It's looking good to us. Over."
102:38:30 **Armstrong:** "It's a 1202."
102:38:32 **Aldrin:** "1202..."
102:38:42 **Armstrong:** "What is it... give us a reading on the 1202 Program Alarm."

As Mission Control puzzled over the alarm reading, *Eagle* passed 33,000 feet (10,000 meters). With just eight minutes until touchdown, they needed to figure out what was going on. Neither Armstrong nor Aldrin had ever encountered a 1202 alarm in training, and both feared that it might endanger the mission and lead to an abort... or worse.

102:38:53 **CapCom:** "Roger. We got you. We're go on that alarm."
102:38:59 **Armstrong:** "Roger."

On the ground, Steve Bales, a flight controller, ascertained that the alarm was a computer overload alert... the primitive guidance computer for the Lunar Module was swamped with data and was protesting. But it was not yet a danger to the landing.

102:42:08 **CapCom:** "Roger. Copy. *Eagle*, Houston. You're go for landing. Over."
102:42:17 **Aldrin:** "Roger. Understand. Go for landing. 3,000 feet. Program Alarm."
102:42:22 **CapCom:** "Copy."
102:42:23 **Aldrin:** "1201..."
102:42:24 **Armstrong:** "1201. Okay, 2,000 [feet

altitude] at 50 [feet per second vertical velocity]."

Armstrong would soon concentrate all his attention on searching for a landing spot, and Aldrin took over calling out the data.

102:42:25 **CapCom:** "Roger. 1201 alarm. We're go. Same type [of alarm]. We're go."
102:42:31 **Aldrin:** "2,000 feet..."

The LM computer had flashed a more urgent warning: a 1201 alarm. This was more than just a protest – it was a threat to quit and leave the astronauts without guidance assistance. It was now a distinct possibility that the mission could be aborted. If this occurred, Armstrong would have to throttle-up, cut loose the LM's descent stage and fire his ascent rockets for an emergency return to orbit. While risky at any altitude, it was increasingly dangerous as they neared the surface, which was now just 2,000 feet (600 meters) below.

Armstrong later commented:
"Our attention was directed toward clearing the program alarms, keeping the machine flying, and assuring ourselves that control was adequate to continue without requiring an abort. Most of the attention was directed inside the cockpit during this time period and... it wasn't until we got below 2,000 feet that we were actually able to look out and view the landing area..."

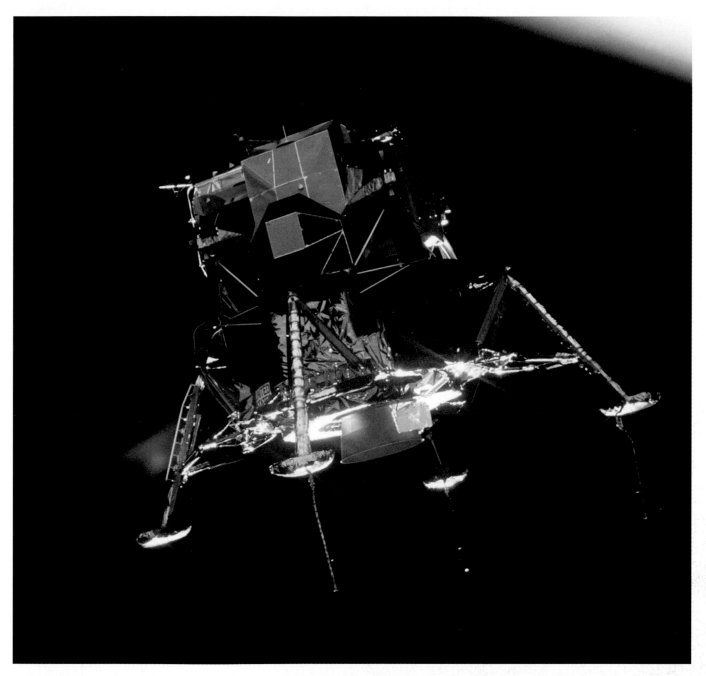

The LM *Eagle* as seen from CM *Columbia*. The rods protruding from the landing pads are the sensors for the contact light, which indicated to the LM pilot that the craft had reached the Moon's surface.

As the computer continued to refuse data, Aldrin glued his eyes to the read-outs.

102:43:01 **Aldrin:** "35 degrees. 35 degrees. 750 [feet]. Coming down at 23."
102:43:07 **Armstrong:** "Okay."
102:43:07 **Aldrin:** "700 feet. 21 down, 33 degrees."
102:43:10 **Armstrong:** "Pretty rocky area."
102:43:11 **Aldrin:** "600 feet, down at 19."

At this point, Armstrong took over manual control of the craft. The skills gained during his years of flying exotic machines such as the X-15 now came into play.

102:43:16 **Aldrin:** "540 feet, down at... 30. Down at 15."
102:43:42 **Aldrin:** "Okay, you're pegged on horizontal velocity."

The LM was now hovering above the surface of the Moon like a helicopter, as Armstrong desperately searched for a landing site. The computer had guided them to an area covered with truck-sized boulders where the surface was supposed to have been clear.

Fuel was now becoming a problem.

102:43:57 **Armstrong:** "270 [feet]. Okay, how's the fuel?"
102:44:00 **Aldrin:** "Take it down!"
102:44:02 **Armstrong:** "Okay. Here's a... looks like a good area here."

Aldrin allowed himself his first look out of the window.

102:44:04 **Aldrin:** "I got the shadow out there."
102:44:29 **Armstrong:** "I got a good spot."
102:44:45 **Aldrin:** "100 feet. Three-and-a-half down, nine forward. Five per cent. Quantity light."

This call by Aldrin caused almost as much sweat on the ground as it did in the LM. This meant that they had only five per cent of their fuel remaining and the craft had to either land in the next 90 seconds or abort. If they attempted to abort at this altitude, the LM would almost certainly crash.

102:44:54 **Aldrin:** "Okay. 75 feet. And it's looking good. Down a half, six forward."

102:45:02 **CapCom:** "60 seconds."

There was now less than a minute left to find a place to land...

102:45:17 **Aldrin:** "40 feet, down two-and-a-half. Picking up some dust."

Armstrong later recalled:

"I first noticed that we were, in fact, disturbing the dust on the surface when we were something less than 100 feet... as we got lower, the visibility continued to decrease... you were seeing a lot of moving dust that you had to look through to pick up the stationary rocks and base your... decisions on that. I found that to be quite difficult..."

The voice from Earth was calm, but firm.

102:45:31 **CapCom:** "30 seconds!"
102:45:32 **Aldrin:** "Drifting forward just a little bit; that's good."
102:45:40 **Aldrin:** "Contact light."

The five-foot (1.52-meter) metal sensor rods that extended from the LM's legs had touched lunar soil. *Eagle* was down. The abort had been avoided by just 17 seconds. It was time to turn off the engine and "safe" the controls...

102:45:43 **Armstrong:** "Shutdown."
102:45:44 **Aldrin:** "Okay. Engine Stop."
102:45:57 **CapCom:**"We copy you down, *Eagle*."
102:45:58 **Armstrong:** "Engine arm is off. Houston, Tranquility Base here. The *Eagle* has landed."

Jubilation rang throughout Mission Control. Gene Kranz, the flight director, stood silent for a few moments. He was not so much savouring the moment, as was simply unable to speak. He finally banged his fist on his console to shock himself into action. It took him a few minutes to restore order, but it was understandable. After eight years and billions of dollars of effort, Apollo was on the Moon.

First Contact

A "Stay/No Stay" determination had to be made. Kranz polled his controllers and soon *Eagle* received approval

NASA and Manned Spacecraft Center (MSC) officials join the flight controllers in celebrating the conclusion of the Apollo 11 mission. From the left, foreground: Dr Maxime Faget, Director of Engineering and Development; George Trimble, Deputy Director; Dr Christopher Kraft Jr, Director of Flight Operations; Julian Scheer (at the back), Assistant Administrator, Office of Public Affairs, NASA HQ; George Low, Manager, Apollo Spacecraft Program, MSC; Dr Robert Gilruth, MSC Director; and Charles Mathews, Deputy Associate Administrator, Office of Manned Space Flight, NASA HQ.

from Mission Control to remain on the Moon. Armstrong and Aldrin began the lengthy preparation to leave the LM and begin humankind's first exploration of another world. First, they had to perform the extended process of making sure the LM was fit to stay on the Moon and return them to Earth. Nobody wanted to discover hours later that they could not come home.

About six hours after touchdown, it was time for Armstrong and Aldrin to begin the exploration of their small corner of the Sea of Tranquility. They had trouble venting enough air from the inside of the LM to open the hatch, which was hinged to swing inward. Although the pressure gauge read zero, it still would not open. Finally, Aldrin reached down to the edge of the thin hatch and peeled it back just a hair's breadth. The remaining air hissed out into the lunar void, creating a halo of ice crystals as it did so. The hatch swung open and Armstrong positioned himself to exit their tiny spacecraft.

109:15:45 **Aldrin:** "Okay. About ready to go down and get some Moon rock?"
109:15:47 **Armstrong:** "My antenna's out."
109:16:49 **Aldrin:** "Okay... forward and up; now you are clear. Little bit toward me. [Pause] Straight down. To your left a little bit. Plenty of room. [Pause] Okay, you're lined up nicely. Toward me a little bit, down. Okay. Now you're clear... you're lined up on the platform. Put your left foot to the right a little bit. Okay. That's good. Roll left. Good."

109:17:54 **Armstrong:** "Okay. Now I'm going to check ingress here."

At this point, Armstrong had to make sure that he would actually be able to get back inside the LM after their explorations or in case of an emergency. He experimented a little with entering the hatch. The opening was so small, and the space inside the craft so tight, that it was not a simple chore. But, as with all aspects of the flight, they had simulated this so often that the procedure was not a great challenge. Concerns abated, he made his way to the "front porch," the small platform at the top of the ladder.

109:19:16 **Armstrong:** "Okay. Houston, I'm on the porch."
109:19:20 **CapCom:** "Roger, Neil."
109:21:07 **Aldrin:** "Did you get the MESA out?"
109:21:09 **Armstrong:** "I'm going to pull it now."

The MESA was a small platform to the left of the ladder that contained the TV camera and access to a number of items needed for the Extra Vehicular Activity (EVA). The most important thing now was to make sure that the

billion-or-so people on Earth who had access to a TV were able to watch the first tentative steps on the Moon.

109:21:18 **Armstrong:** "Houston, the MESA came down all right."
109:22:06 **CapCom:** "And we're getting a picture on the TV!"
109:22:09 **Aldrin:** "You got a good picture, huh?"
109:22:11 **CapCom:** "There's a great deal of contrast in it; and currently it's upside down on our monitor, but we can make out a fair amount of detail."

The first images of a man on another world were fuzzy, black and white and, as it happened, upside down. But it didn't matter. The assembled minions in NASA, and the rest of the world, were thrilled that they were able to see anything. In a few moments, Houston had the picture corrected, and, while still stark with contrast, it was obvious that man was about to set foot on to the Moon.

109:22:48 **CapCom:** "Okay. Neil, we can see you coming down the ladder now."

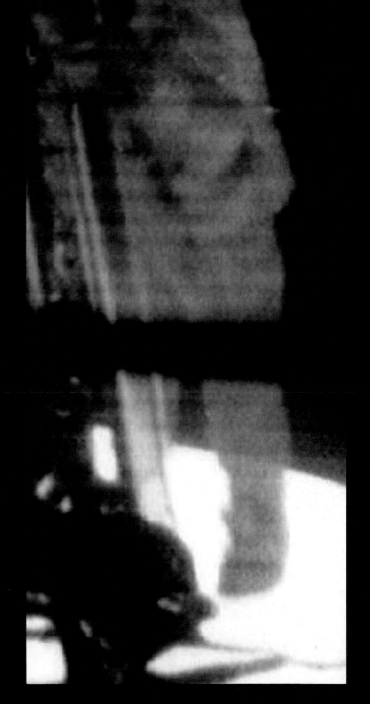

109:22:59 Armstrong: "Okay. I just checked getting back up to that first step..."
109:23:10 CapCom: "Roger. We copy."

The distance from the bottom rung of the ladder to the Moon had been studied almost to distraction. It had to be short enough that a man inside one of the bulky Apollo EVA suits could make the jump, but not so long that it would strike the surface during landing.

109:23:11 Armstrong: "Takes a pretty good little jump... I'm at the foot of the ladder. The LM footpads are only depressed in the surface about one or two inches although the surface appears to be very, very fine grained, as you get close to it. It's almost like a powder. The ground mass is very fine. I'm going to step off the LM now."

Armstrong paused, standing on the LM's large footpad. The dish-shaped pad easily contained his large moon boots. As if to recall the most important sentence of his life, he gathered himself up and put one foot onto the charcoal-gray surface.

"The surface is fine and powdery…" Aldrin's moon boot appears just slightly lifted from the lunar soil. The simple, ridged pattern of the boot's oversized sole is clearly visible in the dust.

109:24:48 **Armstrong:** "That's one small step for man; one giant leap for mankind."

What Armstrong actually said has been famously debated. When considering his first words on the Moon, Armstrong had intended to say: "That's one small step for *a* man, one giant leap for mankind…." Either a communication error occurred, or a mental one. He did not realize the omission until much later, when he exclaimed that "'One small step for man' makes no sense!" Whatever the reason, history remembers the words as they were recorded. Few question the syntax today.

109:25:08 **Armstrong:** "The surface is fine and powdery. I can kick it up loosely with my toe. It does adhere in fine layers, like powdered charcoal, to the sole and sides of my boot. I only go in a small fraction of an inch, maybe an eighth of an inch, but I can see the footprints of my boots and the treads in the fine, sandy particles."
109:25:30 **CapCom:** "Neil, this is Houston. We're copying."
109:25:45 **Armstrong:** "There seems to be no difficulty in moving around as we suspected. It's even perhaps easier than the simulations of one-sixth G that we performed in the various simulations on the ground. It's absolutely no trouble to walk around…"

He took a few more steps.

109:27:13 Armstrong: "Okay. It's quite dark here in the shadow and a little hard for me to see that I have good footing. I'll work my way over into the sunlight here without looking directly into the Sun."

Armstrong later recalled:
"It is very easy to see in the shadows after you adapt for a little while. When you first come down the ladder, you're in the shadow. You can see everything perfectly; the LM and things on the ground. When you walk out into the sunlight and then back into the shadow, it takes a while to adapt."

There had been concerns over every element of these first steps, and much of it had been over a man's ability to see on the Moon. With the high contrast, and the Sun glaring overhead, unfiltered by any atmosphere, Armstrong was cautiously reporting his every move back to Houston.

109:33:25 Aldrin: "Okay. Going to get the contingency sample there Neil?"
109:33:27 Armstrong: "Right."

Armstrong's first task once safely standing on the Moon was to grab some soil, and hopefully a rock, and stow it inside a pocket of his suit reserved for this purpose. This way, if they did have to depart in a hurry, at least NASA would have the consolation prize of a piece of the Moon to look at. As it turned out, they collected nearly 50 pounds (22 kilos) of rocks to puzzle over.

109:34:09 Aldrin: "Looks like it's a little difficult to dig through the initial crust..."
109:34:12 Armstrong: "This is very interesting. It's a very soft surface, but here and there where I plug with the contingency sample collector, I run into a very hard surface. But it appears to be a very cohesive material of the same sort. I'll try to get a rock in here. Just a couple."

Armstrong paused for a moment.

109:34:56 Armstrong: "It has a stark beauty all its own. It's like much of the high desert of the United States. It's different, but it's very pretty out here."

Aldrin was having a hard time waiting calmly inside the LM. He wanted to join his crew member on the lunar surface.

109:38:41 Aldrin: "Okay. Are you ready for me to come out?"
109:39:11 Armstrong: "All set."

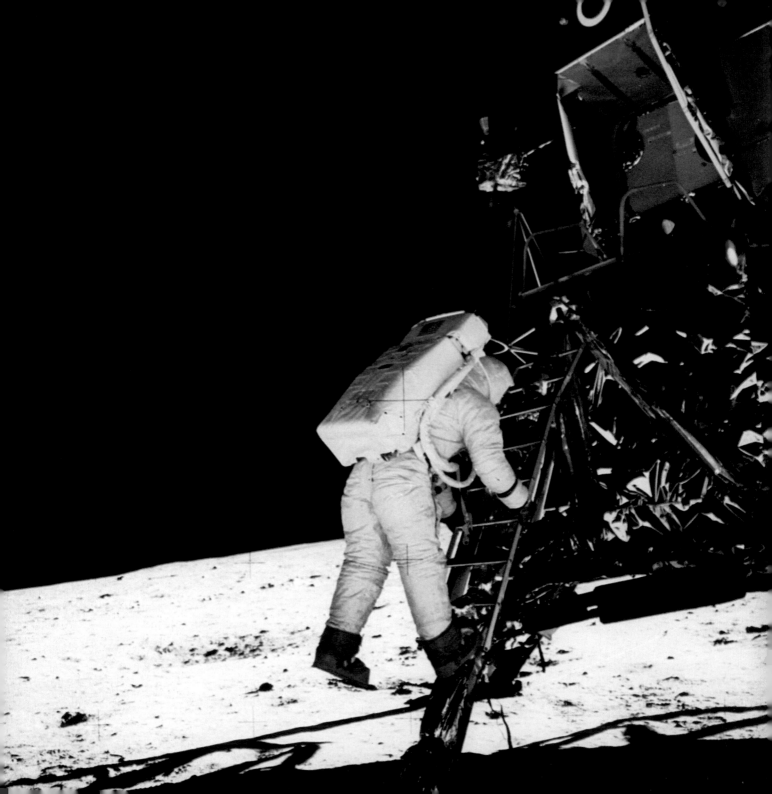

Opposite: Buzz Aldrin standing on the bottom rung of the ladder, ready to hop down to the footpad, as Armstrong had done before him.

Left: The Apollo 11 Moon plaque is seen affixed to the LM landing leg just behind the ladder. It depicts the two hemispheres of the Earth and bears a proclamation that was read by astronaut Neil Armstrong and beamed back to Earth. All subsequent flights carried similar plaques.

Armstrong guided his comrade out of the hatch and down the ladder. Aldrin stepped onto the Moon. He was transfixed.

109:43:16 **Aldrin:** "Beautiful view!"
109:43:18 **Armstrong:** "Isn't that something! Magnificent sight out here."

Aldrin's next words have, for many, summed up the character of the Moon perfectly.

109:43:24 **Aldrin:** "Magnificent desolation."

Aldrin sounded reflective and philosophical. Armstrong, in contrast, sounded almost giddy.

109:52:24 **Aldrin:** "Neil is now unveiling the plaque that is [on the landing] gear."

There were moments on the EVA timeline that were not scientific in nature. One of them was the uncovering and reading of a plaque affixed to the LM's descent stage, which would remain behind when the explorers departed.

109:52:40 **Armstrong:** "For those who haven't read the plaque, we'll read the plaque that's on the front landing gear of this LM. First there's two hemispheres, one showing each of the two hemispheres of the Earth. Underneath it says 'Here Men from the planet Earth first set foot upon the Moon, July 1969 A.D. We came in peace for all mankind.' It has the crew members' signatures and the signature of the President of the United States."

It was official. America had claimed the Moon, and had done so for all humanity.

Down to Work

As Armstrong and Aldrin set about their tasks, Mike Collins checked in from high above them in lunar orbit.

110:08:53 **Collins:** "Houston, *Columbia* on the high gain. Over."
110:08:55 **CapCom:** "*Columbia*, this is Houston. Reading you loud and clear. Over."
110:09:03 **Collins:** "Yeah. Reading you loud and clear. How's it going?"

110:09:05 **CapCom:** "Roger. The EVA is progressing beautifully. I believe they are setting up the flag now."
110:09:14 **Collins:** "Great!"

As the men struggled to erect the flag, which was held aloft by a wire to simulate waving in the wind, Houston narrated the event for Collins.

110:09:43 **CapCom:** "Yes, indeed. They've got the flag up now and you can see the Stars and Stripes on the lunar surface."
110:09:50 **Collins:** "Beautiful. Just beautiful."

At this point, Houston asked Armstrong and Aldrin to move into the camera's view. They were about to receive an important phone call.

110:16:03 **CapCom:** "Neil and Buzz, the President of the United States is in his office now and would like to say a few words to you. Over."
110:16:23 **Armstrong:** "That would be an honor."
110:16:25 **CapCom:** "All right. Go ahead, Mr. President. This is Houston. Out."

110:16:30 **Nixon:** "Hello, Neil and Buzz. I'm talking to you by telephone from the Oval Room at the White House, and this certainly has to be the most historic telephone call ever made. I just can't tell you how proud we all are of what you have done. For every American, this has to be the proudest day of our lives. And for people all over the world, I am sure they, too, join with Americans in recognizing what an immense feat this is. Because of what you have done, the heavens have become a part of man's world. And as you talk to us from the Sea of Tranquility, it inspires us to redouble our efforts to bring peace and tranquillity to Earth. For one priceless moment in the whole history of man, all the people on this Earth are truly one; one in their pride in what you have done, and one in our prayers that you will return safely to Earth."
110:17:44 **Armstrong:** "Thank you, Mr. President. It's a great honor and privilege for us to be here representing not only the United States but men of peace of all nations, and with interests and the curiosity and with the vision for the future. It's an honor for us to be able to participate here today."

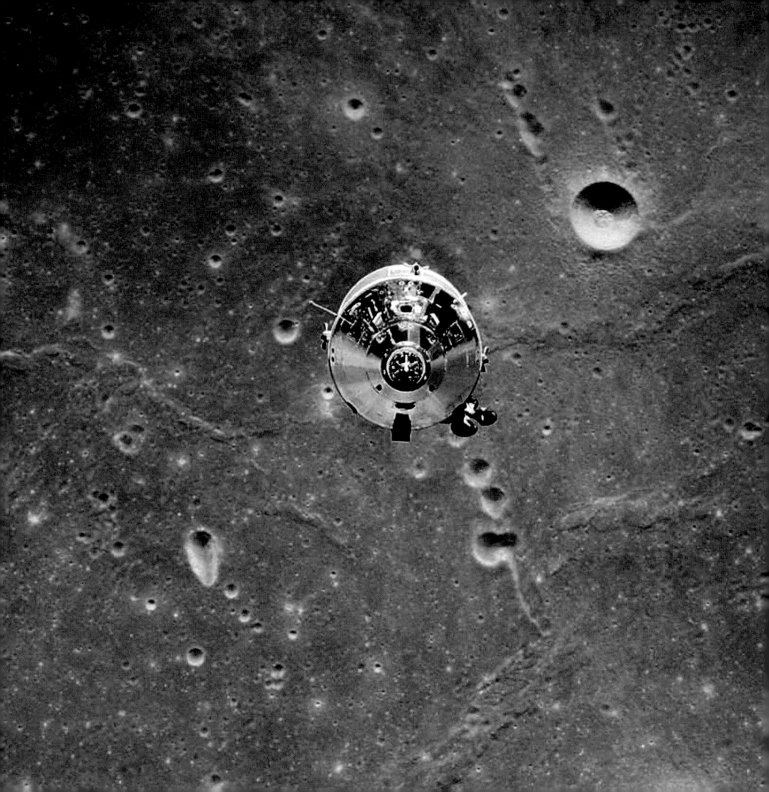

The Apollo 11 CM *Columbia*, as seen from the LM *Eagle*. Michael Collins spent over 24 hours alone in orbit around the Moon during the Apollo 11 mission. Half of each orbit he endured was behind the Moon, and thus out of communication with either his comrades below or the Earth. He later commented, "I knew I was alone in a way that no Earthling has ever been before."

110:18:12 **Nixon:** "And thank you very much and I look forward... all of us look forward to seeing you on the *Hornet* on Thursday."
110:18:21 **Aldrin:** "I look forward to that very much, sir."

The next major item on the checklist was to set up the Early Apollo Surface Experiment Package (EASEP). This was an ingenious array of scientific experiments and measurements, from a seismometer to measure "Moonquakes" to a laser-reflecting mirror box, which would allow scientists to measure the Moon's distance from Earth accurately.

Setting up the science package was a mixed bag. Some of the devices were bulky, and the astronauts had trouble leveling the instruments. Worse still, time was running short.

111:11:15 **CapCom:** "Buzz, this is Houston. You've got about 10 minutes left now prior to commencing your EVA termination activities. Over."
111:11:31 **Aldrin:** "Roger. I understand."

One last important science task required completion: core samples were needed. What lay beneath the Moon's dusty surface was even more important than what lay on top of it. All the while, Armstrong and Aldrin were fighting against the clock.

111:16:13 **CapCom:** "Neil, this is Houston. We'd like you all to get two core tubes and the solar wind experiment; two core tubes and the solar wind. Over."
111:16:25 **Armstrong:** "Roger."

Aldrin moved off to get another core sample.

Armstrong later recalled:
"The primary difficulty that we observed was that there was just far too little time to do the variety of things that we would have liked to have done. We have the problem of a five-year-old boy in a candy store. There are just too many interesting things to do."

Aldrin using his geology hammer to pound a core sample tube into the lunar soil. Behind the core tube, the solar wind collection device, which was essentially a piece of mylar supported by a frame, can be seen.

Aldrin sets up part of the EASEP package. This device is the lunar seismometer, which would measure "Moonquakes" after the two astronauts had departed.

Both astronauts regretted not having more time to collect and properly label rock samples. But it would have to suffice. The two men, tired from working in their clumsy suits, closed up their experiments and samples, and climbed back into *Eagle*. They were still exuberant, but also feeling somewhat distressed over the lack of time for all their tasks. This frustration led to timeline changes for future missions.

Armstrong recalled:

"We did get almost everything done that we had intended to do. Some things didn't go as smoothly as we would have liked, but we also did some things that we hadn't expected to do. On balance, I had to be pretty pleased with the way it came out."

Back in the LM, nearing liftoff, Aldrin suddenly thought about the first samples of lunar soil Neil had taken – the contingency sample. An almost comic discussion of the first piece of the Moon in human possession followed.

Aldrin: [To Neil] "Uh – where's the contingency sample right now?"

Armstrong: "I don't know. I mean, we don't care about that any more, once we got the rock boxes."
Aldrin: "Did we throw it out? Or did we keep it?"
Armstrong: "Oh, I think we kept it."
Aldrin: "How did it get back without kind of being out loose and all that?"
Armstrong: "It was in my pocket."
Aldrin: "I'm just trying to... because I don't remember dealing with it hardly at all."
Armstrong: [Chuckling] "It may still be in that suit pocket, for all I know." [Referring to the moonsuit he had just thrown outside the LM door.]
Aldrin: "You got it at home in your top drawer?"

For the record, history reports that the contingency sample did make it back into the LM, along with 46 pounds (21 kilos) of rocks and soil. Their job was done. By the time newspapers back on Earth had printed up and distributed their special editions, the two men had rejoined their colleague in orbit and were heading home to Earth. Man's first exploration of another world was now a part of the history books, and our world would never be quite the same.

07 APOLLO 12:
THE LIGHTER SIDE OF SPACE

November 1969: Oceanus Procellarum

Just four months after the first Moon landing, another crew landed on the rocky surface. This time, they were instructed to set down close to the lunar probe Surveyor III, which landed there three years earlier.

The second landing on the Moon was halfway across its face from the Sea of Tranquility in the Ocean of Storms (*Oceanus Procellarum*). Once again, a tiny speck of flame illuminated the empty skies around the Moon. The flame grew until another ungainly Lunar Module, this time named *Intrepid*, rested quietly on the surface of Earth's nearest neighbor. But that is where any similarity to the first landing ended. For although the two men inside were charged with a similar mission of science and exploration, the tone of their flight was entirely different. Their names were Pete Conrad and Alan Bean.

Dr. Thomas Paine, NASA Administrator, shields the First Lady, Mrs. Richard M. Nixon, from rain while the President and their daughter Tricia (foreground) watch the Apollo 12 countdown.

115:04:15 **Bean:** "Those rocks have been waiting four-and-a-half billion years for us to come grab them."
115:04:20 **Conrad:** "Think so, huh?"
115:04:21 **Bean:** [Laughing] "Let's go grab a few."

In stark contrast to the tension of the Apollo 11 mission, Pete Conrad, Alan Bean and Richard "Dick" Gordon laughed their way to the Moon. Dry jokes and bad puns were the order of the day. The astronauts had the time of their lives on their mission, and they let everyone listening know it.

Down on the surface, Conrad and Bean were soon ready to make their way outside from *Intrepid*, but Mission Control had some housekeeping to do first, and told the two to wait. Conrad became impatient.

Crew: Apollo 12

Charles "Pete" Conrad's intense intelligence was belied by his wry, wisecracking sense of humor. His way with words, which was of concern to more than one person in NASA's public relations department, was often spectacularly vulgar. He was born in Philadelphia, Pennsylvania, in 1930, and earned a degree in aeronautical engineering from Princeton University in 1953. He soon entered the US Navy, and, after a spell as a test pilot, he joined NASA just in time for the Gemini missions. Besides his easygoing humor and ready smile, Conrad, who commanded Apollo 12, is credited with forging the closest crew in the program.

Alan Bean, known to his comrades as "Beano," was born in 1932 in Texas. He received a Bachelor of Science degree in Aeronautical Engineering from the University of Texas in 1955. Before becoming an astronaut, Bean was a US Navy pilot, flying over 5,500 hours in 27 different aircraft. In 1963 he was admitted to the ranks of NASA, where his soft-spoken intelligence won him an early seat to the Moon.

The third member of the Apollo 12 team was Richard Gordon, better known as Dick. He was born in 1929 in Seattle and held a Bachelor of Science in Chemistry from the University of Washington. After earning his aviator wings in the US Navy, Gordon was chosen for the third intake of astronauts in 1963. He flew on Gemini XI in 1966 and was the backup Command Module pilot for Apollo 9. Now, as Command Module pilot for Apollo 12, he would maintain the lonely vigil in orbit until his friends returned.

Pete Conrad (left), Dick Gordon (center) and Alan Bean (right) strike a characteristically jolly pose during a visit to North American Rockwell Space Division in Downey, California, for a spacecraft checkout.

115:05:15 **Conrad:** "'Stand by?' You guys ought to be spring-loaded!"

Finally, Conrad and Bean were given permission to go outside. But they had trouble with the hatch on their LM, just as Armstrong and Aldrin had. Bean reached down and pulled on the edge of the hatch to let the remaining air escape. Pete Conrad later recalled the hatch opening:

"That hatch had some ribbing that you could get a hold of. You realize this whole vehicle was pretty tinny. So this thing was very easy to [open]... He could actually pull a little corner open."

Conrad started down the ladder.

115:18:37 **Conrad:** "Hey, I'll tell you what we're parked next to..."
115:18:39 **Bean:** "What?"
115:18:40 **Conrad:** "We're about 25 feet in front of the Surveyor Crater!"

Conrad was ecstatic. After the far-downrange landing of Apollo 11, one of his personal aims was to land his craft much closer to the target than Armstrong had. He came as close as anyone had a right to expect, within about 100 yards (91 meters). Only a bit farther away was the Surveyor III unmanned probe, which had landed three years earlier. It had been baking in the lunar sun for all this time, and NASA wanted a piece of it back. Many doubted that Apollo 12 could land close enough to

Pete Conrad's pinpoint landing of *Intrepid* on Apollo 12 confirmed that the LM could be guided to a precise location. The crew were a quick walk from their target, the Surveyor III probe, which had landed there three years earlier.

Crews on the first two landings made detailed examinations of the effect of the landing rocket on the lunar surface. Here some scouring of the soil can be seen, but it was less than had been expected. The metal rod on the right is the contact probe rod, which dangled from the footpads in flight. When it touched the Moon's surface it triggered a blue "contact light" inside the LM to let the pilot know that he had landed.

from San Francisco, he carried it off with little drama.

Soon Conrad had reached the last rung on the ladder. At five feet six inches (1.67 meters) his short stature, in comparison to the six-foot (1.8-meters) Armstrong, made for a classic remark upon his arrival to luna firma.

115:22:16 Conrad: "Whoopee! Man, that may have been a small one for Neil, but that's a long one for me."

Once off the footpad he could see the object of his desire.

115:23:27 Conrad: [Gleeful] "Boy, you'll never believe it. Guess what I see sitting on the side of the crater!"
115:23:30 Bean: "The old Surveyor, right?"
115:23:31 Conrad: "The old Surveyor. Yes, sir. Does that look neat! It can't be any further than 600 feet from here. How about that?"
115:23:43 CapCom: "Well planned, Pete."

Once Conrad got to the surface he began to deploy tools and experiments from the MESA on the side of the LM landing stage. As Al Bean came down the ladder, Pete indulged in a favorite pastime. He loved to hum aimless tunes – it drove the ground controllers crazy.

115:38:55 **Conrad:** "... Bo bo bo bo; dee dee; dee dee dee. I tell you one thing, we're going to be a couple of dirty boogers [referring to the sticky, dark lunar soil that would soon coat their spacesuits]... I tell you, this is dirt dirt."

As Bean readied to leave the LM, Conrad prepared the TV camera for use. It was a color camera, unlike Apollo 11's, and promised to stun the world with wondrous images of the mission.

115:42:00 **Conrad:** "Okay; let me see. While you're doing that, what was I supposed to do? Oh, I know: 'Possible TV deploy.' I'll go work on the tripod."
115:42:15 **Bean:** "Okay."
115:42:17 **Conrad:** "Dum dum, da dee da dee dum. Trying to learn to move faster. Pretty good. Hey, I feel great."

It was time for Bean to set up the TV camera. The portable color camera was very sensitive to bright light, and as Bean was orienting the unit, it caught a strong reflection off the LM and burned out.

115:59:45 **CapCom:** "Al, we have a pretty bright image on the TV."
115:59:46 **Bean:** "Okay."
115:59:47 **CapCom:** "... could you either move or stop it down?"
115:59:52 **Bean:** "Okay, I'm going to have to stop it down. That's as far as it goes, Houston. How does

that look to you?"
116:00:08 **CapCom:** "No, it still looks the same, Al. Why don't you try shifting the scene?"

Bean later recalled the incident:
"I think this is where we really got into our first problem. I took the TV off the MESA pretty readily and stuck it on top of the tripod and moved the tripod and the TV over to the deployment... I carried the camera over to the opposite side, stuck it there, pointed it at the LM, and called the ground. It looked to me like there were some pretty bad reflections off the LM and I was concerned that maybe they'd bother the TV. Apparently, that's just exactly what happened; these reflections were far too bright for the TV to handle and it burned out.

Bean continued to struggle with the suddenly very-important camera.

116:01:01 **Bean:** "Wait a second. I got to... How does that look, Houston?"
116:01:05 **CapCom:** "Still looks the same, Al. We have a very bright image at the top and blacked out for about 80 per cent of the bottom."
116:01:11 **Bean:** "Well, I'll tell you what [you] let me do, Houston. Let me move it around here so the back is to the Sun, and maybe that'll help. Maybe that's the way we're going to have to do it."
116:01:31 **CapCom:** "Okay, Al; go ahead."
116:02:01 **Bean:** "What do you see now?"

The dead TV camera belonging to Apollo 12. As Bean positioned it, the delicate camera caught a glare from the foil-covered LM and burned out. It was the first color camera on the Moon's surface.

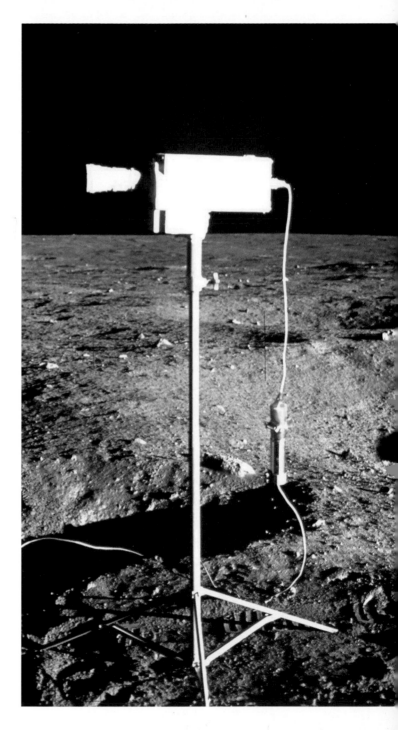

116:02:03 **CapCom:** "Still the same, Al. We've got a very bright part – about 20 per cent of the top – and black on the bottom."
116:02:15 **Bean:** "Well, got any suggestions?"
116:02:19 **CapCom:** "Stand by, Al."

The world stared at a black screen with a white smear across the center. The show was over. Try though they might, in the end it was clear that the camera had burned out. American television networks scrambled to accommodate this new wrinkle. CBS switched to their cameras at Grumman Aerospace, builders of the LM, who had a full-scale lunar surface mock-up with a couple of employees romping about in Apollo suits. They tried their best to follow what the astronauts were doing from the verbal downlink. The effect was passable. ABC was a little more creative. They hung a lunar backdrop and had two men dressed in "spacesuits" that looked for anything like costumes from *The Twilight Zone*.

But the real comedy was on NBC, who had contracted a puppeteer to create Apollo marionettes for simulations. They had a small lunar surface mock-up, and soon two tiny puppets, strings clearly visible, were bouncing their way across the lunar surface.

The Apollo 12 mission continued as a radio-only event. After giving up on the TV camera, they moved on to other tasks, including the deployment of the ALSEP (Apollo Lunar Science Experiment Package), the new instrumentation that would stay on the Moon after their departure. It was a sophisticated bundle of

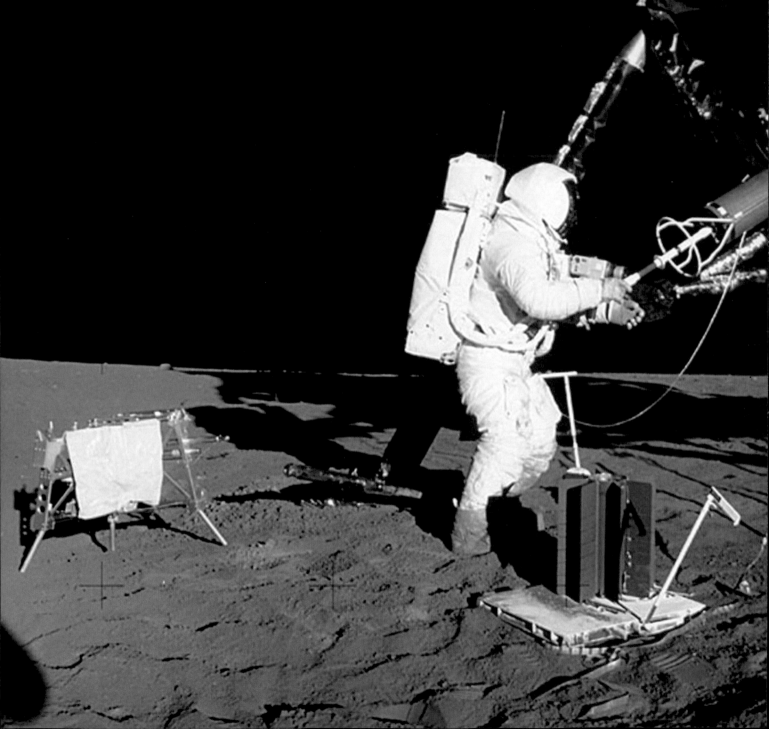

Al Bean attempting to remove the plutonium fuel rod from its holder in the LM. However, its tolerances had changed during the flight, and it would not budge. The RTG generator, in which it was to be inserted, lies next to his feet. The object to the left is the tool carrier, in which the astronauts carried most of the implements needed for their explorations.

science, powered by a nuclear fuel source. They prepared to remove the Radioisotope Thermal Generator (RTG) fuel rod from its protective covering inside the LM descent stage.

116:40: **Bean:** "Houston, we're going to go ahead and pull down the fuel cask right now, and then I'll take the element out of it."

The two men collaborated on the task.

116:40:26 **CapCom:** "Roger, Al. Copy. You're working with the fuel cask."
116:40:31 **Conrad:** "Wait, wait, wait, wait, wait!" [Laughs]
116:40:39 **Bean:** "That's a bad place to put it."
116:40:40 **Conrad:** "Yeah."

The fuel rod was stuck. It wouldn't budge. The carbon container in the LM's descent stage was designed to contain the plutonium inside in case of an abort in the Earth's atmosphere. Otherwise, when the LM broke up during re-entry, the deadly element could have been

Conrad retrieves the Surveyor camera, which had spent three years on the surface of the Moon.

spread over a wide area. Once on the Moon, the fuel had to be removed from the cask and placed inside the RTG generator, which powered the ALSEP package. Conrad told Houston that the delicate retaining mechanism for the fuel rod was stuck.

116:45:13 **Conrad:** "[You] guys got any suggestions?"
116:45:17 **Bean:** "It really kind of surprises me."
116:45:25 **Conrad:** "Come over and look."

With characteristic aplomb, Bean had an idea.

116:46:50 **Bean:** "... go get that hammer and bang on the side of it."

116:46:52 **Conrad:** "No. I got a better idea. Where's the hammer?"
116:47:06 **Bean:** "[The] hammer's on the MESA."
116:47:08 **Conrad:** "Okay."

Pete proceeds to bang the side of the fuel cask with the hammer.

116:48:56 **Bean:** "Hey, that's doing it! Give it a few more pounds. Got to beat harder than that. Keep going. It's coming out. It's coming out! Pound harder."
116:49:08 **Conrad:** "Keep going."
116:49:10 **Bean:** "Come on, Conrad!"
116:49:14 **Conrad:** "Keep going, baby."

A closer view of Surveyor with *Intrepid* in the background. The two panels on top of the probe are solar panels to make electricity. The scissor-like extension on the right is the digging scoop, and the vertical tubular structure above and left of it is the camera, which the astronauts retrieved.

116:49:15 **Bean**: "That hammer's a universal tool."

116:49:17 **Conrad**: "You better believe it…"

116:49:18 **Bean**: "There, you got it!"

116:49:19 **Conrad**: "Got it."

116:49:20 **Bean**: "Got it, Houston. That's beautiful. That's too much."

116:49:24 **CapCom**: "Well done, troops."

They inserted the fuel rod into the ALSEP generator and moved on.

After some rock sampling and nearby exploration, the two men climbed back into their LM, tired and dirty, to rest and prepare for their next EVA to Surveyor III.

Shake Hands with the Robot

Midway through their second EVA Conrad and Bean made their way to the edge of nearby Surveyor Crater. Downslope and far ahead, the Surveyor III probe sat as it had since landing three years before. The two marveled at the sight, and then headed down the slope. They had instructions to examine Surveyor and remove some components for return to Earth.

As they neared the robotic probe, Bean noticed where the soil sample scoop it carried had dug into the lunar soil 31 months before. On the airless Moon, little had changed since the Surveyor's arrival.

After a brief examination, Conrad was supposed

to cut some samples from the Surveyor. Bean retrieved a pair of ultra-lightweight bolt-cutters from Conrad's suit. As Conrad prepared to salvage the machine, he paused.

134:20:33 **Conrad:** "Son of a gun."
134:20:33 **Bean:** "Uh – a few extras."
134:20:35 **Conrad:** "It is. It's wired entirely different."

The spacecraft that sat before them was very different from the supposed twin they had trained on back in Houston. Conrad had to make some choices about what to cut for retrieval, then began the dismemberment. The support struts for the onboard camera, one of their target objects, were more formidable than Conrad had anticipated. He grunted as he fought to cut through the metal rods.

134:21:17 **Conrad:** "They've got to be kidding."
134:21:30 **Bean:** "Got it? No, it's too big… "
134:21:33 **Conrad:** "Huh? Too big? Wait, wait, wait, wait, wait!"

Conrad continued to grind away at the support tubes and eventually freed the camera.

Within 24 hours, Conrad and Bean had left the Moon, docked with the CM *Yankee Clipper* and rejoined a very happy Dick Gordon. Gordon had been busy with his own schedule of photography and surface mapping while the other two had been frolicking on the Moon's surface. Still, he was very glad to see them. Behind the Moon for the last time, they jettisoned the LM ascent stage and fired the rocket engine on the Service Module to send them home.

Re-entry and splashdown were as smooth as glass. And while they went their separate ways after the flight of Apollo 12, the bond between the crew remained, and they were close friends for life. They were also some of the happiest astronauts, and their feelings about life might be best summed up by the words of Pete Conrad when Houston told them they were to go to the Moon on Apollo 12: "Whoop-de-doo! We're ready. We didn't expect anything else!"

06 APOLLO 13 – STRANDED

55:54:02 Mission Elapsed Time
200,000 miles (321,800 kilometers) from Earth

After the fire on board Apollo 1, the Apollo 13 mission was the closest NASA came to a major disaster in the lunar landing program. Far into its outbound journey, an onboard tank exploded in the Service Module, causing the crew to lose critical supplies of oxygen, water and power. Their only chance of survival was to use the Lunar Module as a lifeboat, a task for which it had never been designed.

"Bang!"

This was probably the sound an astronaut least wanted to hear 200,000 miles (321,800 kilometers) from Earth. As it reverberated through the Command and Lunar Modules, Jim Lovell, Apollo 13's commander, looked angrily at crew mate Fred Haise. The playful Haise had a habit of twisting a re-pressurization valve in the LM and when it thumped, it always shook up the other guys. But the look on Haise's face told Lovell that this was no joke. Moments later, Jack Swigert, the Command Module pilot, saw the instrument panel before him light up like a Christmas tree.

Alan Shepard, who was America's first man in space in May 1961, and who later flew on Apollo 14, looks on as controllers struggle to save his comrades on Apollo 13

Swigert: "Okay, Houston, we've had a problem here."
CapCom: "This is Houston. Say again please."

By now Lovell had drifted from the attached LM back to his seat in the CM and saw, to his horror, what Swigert was looking at.

Lovell: "Ah, Houston, we've had a problem."

Both men stared at the confusion of warning lights and gauges, trying to make sense out of chaos.

Lovell: "We've had a main B bus undervolt."

Back on the ground at Mission Control, everyone

Flight controllers keep a close watch on Apollo 13.

snapped to rigid attention. Flight Director Gene Kranz looked over the heads of his controllers, and began to poll them over the Flight Director's audio loop.

CapCom: "Roger. Main B undervolt. Okay stand by 13, we're looking at it."

The crew was in no mood to "stand-by." Lovell thought that a meteorite might have hit the craft. Within minutes, with no pressure alarms, he knew that was not the case. But on the ground it was clear that things were getting worse, not better, and it was becoming apparent that this was not an instrumentation problem. Something real had gone wrong – something big.

Back in the dying CM, *Odyssey*, the crew were

rapidly losing power and life support. As concern mounted in Houston, Lovell noticed a smoky haze outside the window. There should not be haze in space. There could be only one conclusion.

Lovell: "We are, we are venting something out... into space."
CapCom: "Roger. We copy you're venting."

In Houston, Kranz still needed answers – and he needed them immediately.

Kranz: "Okay let's everybody think of the kind of things we'd be venting. GNC, you got anything that looks abnormal in your system?"
GNC: "Negative, Flight."

Gene Kranz, the Flight Director whose calm leadership under fire has frequently been credited with saving the crew of Apollo 13.

Kranz: "How about you, EECom, see anything that, with the instrumentation you got, that could be venting?"
EECom: "Let me look at the system... as far as venting is concerned."

Survival

Apollo 13 was in deep trouble. It was not long before the explosion of oxygen in tank number two cascaded into a series of misfortunes that threatened the lives of the crew and tested NASA's ingenuity and the astronauts' abilities to their limits.

The accident had its origins in an older Apollo design. Each oxygen tank in the Service Module had a small heater with a thermostat element inside. They had originally operated at 28 volts. NASA later changed this specification to 65 volts, but somehow the tank in Apollo 13 never got updated. When the high voltage ran through it, it shorted, sparked and – packed with pure oxygen – exploded.

On the ground, Kranz was preparing his control team for an alternative mission. He gazed across the gathered faces, all of which were silent.

Kranz: "Okay, now let's everybody keep cool. We got the LM still attached. The LM spacecraft's good so if we need to get back home we've got a LM to do a portion of it with... We'd like to have RCS (Reaction Control System), but we got the Command Module system so we're in good shape if we need to get home. Let's solve the problem but let's not make it any worse by guessing."

Crew: Apollo 13

Commander James Lovell was born in Cleveland, Ohio, in 1928. He was older and more experienced than many of his fellow astronauts. After attending the University of Wisconsin, from which he graduated in 1952, he went on to test pilot school in Maryland. Joining NASA in 1962, he flew on Gemini IX and Apollo 8. Lovell was no stranger to space travel, and his vast experience was tested to the limit during the Apollo 13 mission.

Fred Haise was younger than Lovell, having been born in Mississippi in 1933. He earned his Bachelor of Science degree with honors in Aeronautical Engineering from the University of Oklahoma in 1959. He spent several years as a test pilot for NASA before becoming an astronaut in 1966. While he had served as a backup crew member for Apollo 8 and 11, this was his first spaceflight. It would be one for the books.

John "Jack" Swigert wasn't supposed to fly on the mission. His seat had originally been allocated to Ken Mattingly. However, a few weeks before takeoff, Mattingly had been exposed to German measles. Though he showed no outward signs of illness, the flight surgeons were adamant that he should not fly, and Swigert, the backup CM pilot, became a member of the crew. Born in Colorado in 1930, Swigert earned a Bachelor of Science in Mechanical Engineering from the University of Colorado in 1953, and a Master of Science degree in Aerospace Science from the Rensselaer Polytechnic Institute in 1965. Joining the Air Force in 1953 he became a fighter pilot in Japan and Korea. After logging 7,200 hours of flight time, he was selected as an astronaut by NASA in April 1966.

Fred Haise (left), John "Jack" Swigert (center) and James Lovell (right). This photo replaced the original Apollo 13 prime crew photo, which had Ken Mattingly in the center.

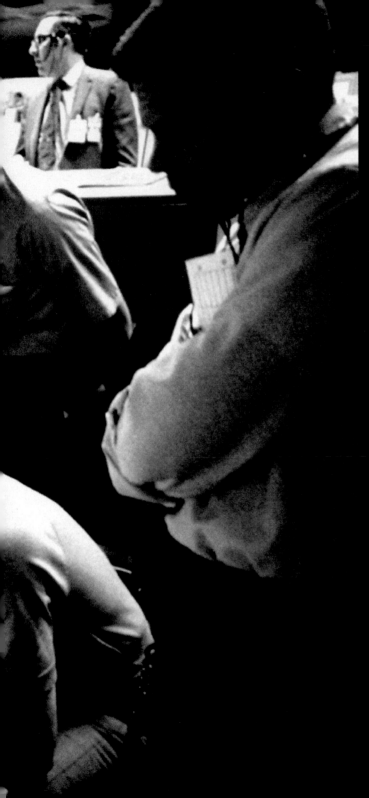

Mission Control works to solve Apollo 13's fuel cell crisis. Featured in frame are eight astronauts and flight controllers. Seated, from right to left are Astronaut Alan Shepard, Jr; Astronaut Edgar Michell; and Guidance Officer Raymond Teague. Standing, from right to left, are M.P. Frank, a flight controller; Astronaut Ronald Evans; Astronaut Eugene Cernan; Astronaut Joe Engle; and Scientist-Astronaut Anthony England.

In an effort to staunch the flow of oxygen that was rapidly departing the spacecraft, Mission Control decided to shut down the fuel cells, from which Apollo 13 should have been drawing critical supplies like power and water.

CapCom: "Okay, 13, this is Houston. It appears to us that we're losing O_2 flow through fuel cell three, so we want you to close the reac valve on fuel cell three. You copy?"
Haise: "Did I hear you right? You want me to shut the reac valve on fuel cell three?"
CapCom: "That's affirmative."

There was a tense moment in the CM as this sunk in. Once the fuel cells were shut down, they could not be restarted. Little did anyone know that this would be the least of the crew's problems in a few hours.

Haise: "You want me to go through the whole smash for fuel cell shut down?"
CapCom: "That's affirmative."

The Public Affairs Officers at John Space Center in Houston (JSC PAO) had been on the air from the time of launch. Suddenly their announcements were being carried worldwide, and as people learned of the emergency, they hung on every word.

Concern at Mission Control. Although there were three flight control teams, each working an eight-hour shift, few went home during the Apollo 13 crisis. Most slept down the hall until it was resolved.

57:11:00 Mission Elapsed Time

JSC PAO: "This is Apollo Control, Houston. Fifty-seven hours, 11 minutes. We now show an altitude of 180,521 nautical miles. Here in Mission Control we're looking... towards an alternate mission, swinging around the Moon and using the Lunar Module power systems, because of the situation that has developed here this evening."

In Mission Control, the reality had set in that the LM was going to be the crew's salvation – or a part of their collective tomb. The mission would take a radically different course to completion. The idea was to alter the craft's trajectory so that instead of entering an orbit around the Moon they would swing behind it and then head back toward Earth. Nobody wanted to risk using the Service Module engine, as it was impossible to tell how the explosion might have affected it. They would have to depend on the LM descent stage to make the corrective burns.

Lovell: "Okay. *Aquarius* is up and *Odyssey* is completely powered down according to the procedures you read to Jack."
CapCom: "Roger, we copy. That's where we want to be, Jim."

In space, Apollo 13 was preparing for a mid-course correction and, once that was completed, shutting down all non-essential systems to conserve power. The LM, designed to hold two men for just over two days, now had to shelter and care for three until the

The Moon as seen from the LM *Aquarius*. As they prepared to go behind the body, Lovell reminded his crew mates that they needed to focus on getting home, to which they replied, "You've been here before… we haven't!" and stole a few more glances out the window.

mission was concluded – 84 hours. It was considered safer to go around the Moon, and once the ship cleared the far side, let gravity do the rest and bring them home. It would require just a few firings of the LM engine to position them for re-entry.

But to make multiple course corrections required navigation, which was becoming yet another problem. They did not know how accurate their computer's bearings were, and had no way of checking. So, like the ancient mariners, the crew of Apollo 13 would be navigating by the stars. In this situation, it seemed anything but quaint.

Into the Night

At 61 hours 30 minutes into the flight, it was time to perform the corrective burn to switch the crew to a "free-return" trajectory. If, for some reason, this did not work, they would still swing past the Moon, but miss the Earth upon returning by some 2,400 miles (3,800 kilometers). Eventually, about six weeks from launch, the three asphyxiated astronauts and their lifeless craft would re-enter Earth's atmosphere and plunge into an ocean. The crew prepared to fire the engine on the LM descent stage.

Haise: "Master arm is on, one minute."
CapCom: "Roger *Aquarius*, and you're go for the burn."

The engine ignited. It burned properly and would do the job. But it was designed to fire only once. Apollo 13 would need to use it multiple times.

Lovell: "Fifty per cent [thrust]."
CapCom: "Okay *Aquarius*, you're looking good."
Lovell: "Auto shutdown."

Success. They headed off to the loneliest of places, the orbital swing behind the Moon. Soon they were out of radio contact.

When they re-emerged it was time for another burn. At this point, their navigational data was suspect. To make this worse, any errors now would simply compound over time.

Deke Slayton, head of the Astronaut Office, explains the proposed "fix" for the CO_2 problem onboard Apollo 13 to Mission Control. They were strictly limited to what materials the astronauts had on board.

79:30:00 Mission Elapsed Time

JSC PAO: "Now one minute away from scheduled time of ignition... less than 30 seconds away... engine is armed. Standing by."
CapCom: "Jim, you are go for the burn, go for the burn."
Lovell: "Roger. Understand go for the burn."
JSC PAO: "Ground confirms ignition."
Lovell: "We're burning 40 per cent."
CapCom: "Houston copies."
Lovell: "Shutdown."
CapCom: "Roger. Shutdown."

79:32:00 Mission Elapsed Time

JSC PAO: "That was Commander Jim Lovell reporting shutdown, engine is off."

After the maneuver, the three men tried to sleep, but in the harsh cold of space it was not easy. It was like camping in a damp trash dumpster in summer clothing. Nobody could get comfortable.

They were soon awake and struggling with a new problem. The lithium hydroxide filters, which both spacecraft used to keep the air breathable, were losing their effectiveness. They needed to change to a new one, and only the LM's unit was working. Unfortunately, they had used up the LM filters, and the CM's units did not fit. Nobody ever thought they would have to. So the engineers in Houston worked out an effective, if inelegant, solution, utilizing materials that the crew had onboard the spacecraft. Lovell and Swigert gathered the needed supplies.

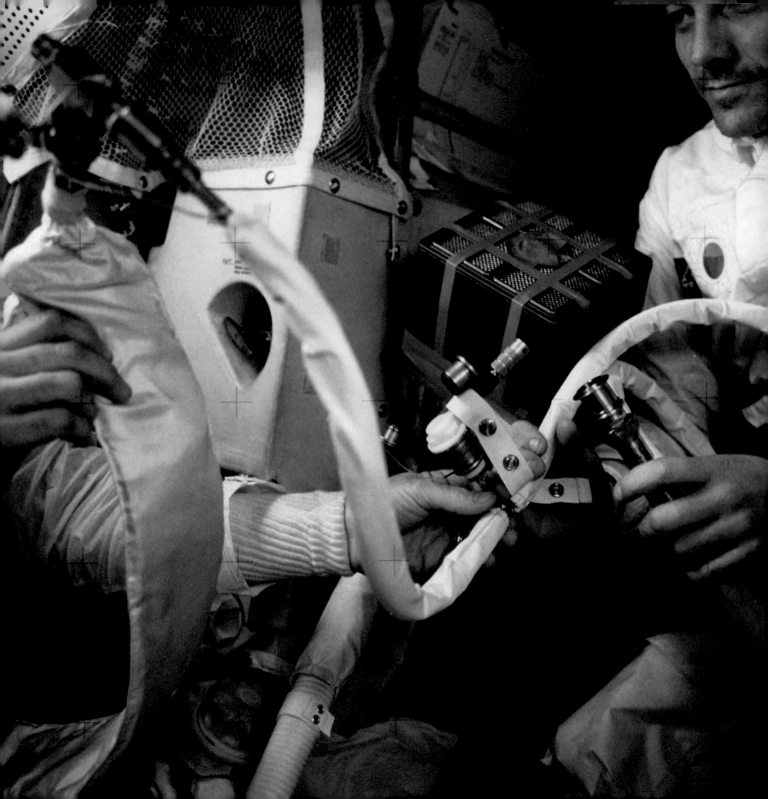

The instructions came up through the radio, step-by-step. It was tedious work, and to the shivering astronauts, it was torture. By the time they were done, Lovell and Swigert were exhausted. But the unorthodox idea had worked, and they lived through the night.

Coming Home

138:01:00 Mission Elapsed Time

After yet another course correction, this time using the Earth as a visual reference, it was time to lighten the spacecraft's load for re-entry. The Command Module was meant to re-enter the atmosphere without the Service Module, which had been virtually useless since the accident anyway. The SM would now be kicked loose.

JSC PAO: "Apollo control Houston, 138 hours, 1 minute into the flight... we now show Apollo 13 at a distance of 35,729 nautical miles away from Earth traveling at a speed of 10,400 feet per second."

The switch was thrown. The three men crowded to the CM windows in an attempt to see what had befallen their ship as the Service Module drifted off. The sight that greeted them was worse than they had imagined.

Lovell: "SM sep."
CapCom: "Copy that."
Lovell: "And there's one whole side of that spacecraft missing!"
CapCom: "Is that right."
Lovell: "Right by the... look out there would you. Right by the high-gain antenna the whole panel is blown out almost from the base to the engine."
CapCom: "Copy that."
Haise: "And it looks like it got to the SPS bell too, Houston."

The SPS bell was the large rocket engine on the Service Module. As it turned out, had they succumbed to the temptation of firing it to return home early, it would likely have exploded.

This image, a frame from a 16mm movie of the CM jettison, shows the entire missing side of Apollo 13's damaged Service Module. The explosion of the oxygen tank did more damage than is visible here.

CapCom: "You can see it dinged the SPS engine bell, huh?"
Swigert: "Man, that's unbelievable."

141:01:00 Mission Elapsed Time.

Tensions had lightened a bit. The major tasks remaining for the crew were to jettison the LM and position the capsule for re-entry. The astronauts began to joke a bit, a sure sign that they were feeling hopeful and were glad to be closer to home.

Lovell: "Well I can't say that this thing hasn't been filled with excitement."
CapCom: "Well, James if you can't take any better care of a spacecraft than that, then we might not give you another one. "

It is not recorded what expression crossed Lovell's face after that crack. As it turned out, he had decided this was his last flight anyway. They were now officially done with the LM *Aquarius*, so with some reluctance, they let her go.

Lovell: "LM jettison."
CapCom: "Okay, copy that."

CapCom: "Farewell *Aquarius*, and we thank you."
JSC PAO: "This is Apollo control, Houston... we've
had Lunar Module jettison. And for Apollo 13, the age
of *Aquarius* ended at 141 hours, 30 minutes ground
elapsed time."

A Fiery Return

141:31:00 Mission Elapsed Time

As they entered the upper reaches of the Earth's
atmosphere, many on the ground were still uncertain if
the astronauts would survive the inferno of re-entry.
What if the heat shield has been cracked by the blast
or by the cold of space? But there was little point in
telling the crew. Instead the astronauts thanked the
flight controllers for their help.

Swigert: "Hey, I want to say you guys are doing real
good work."
CapCom: "So are you guys, Jack."
Swigert: "I know all of us here want to thank all of you
guys down there for the very fine job you did."
CapCom: "I tell you, we all had a good time doing it."
Lovell: "You have a good bedside manner."
CapCom: "That's the nicest thing anybody's ever said."
CapCom: "Okay. LOS [Loss Of Signal] in a minute or a
minute and a half... "
Swigert: "Thank you."

The crew of Apollo 13 plunged into the furnace of re-

entry and lost contact with Houston, as was
expected. The planned three minutes went by – and
the ground heard nothing.

JSC PAO: "Apollo 13 should be out of blackout at this
time. We are standing by for any reports... "

Mission Control was hoping for a visual sighting of the
landing parachutes. Of course, this would not mean
that the crew was alive, only that the capsule had
survived, as the parachute deployment mechanism
was automatic. The minutes ticked by.

JSC PAO: "... standing by for any reports of acquisition."

A tracking aircraft finally called in.

JSC PAO: "We got a report that ARIA 4 aircraft has
acquisition of signal."
CapCom: "Odyssey, Houston. Standing by. Over."

The message swam in a sea of static, but it was all
Mission Control needed to hear.

Swigert: "Okay, Joe."
CapCom: "Okay. We read you, Jack."
JSC PAO: "That was... Command Module pilot Jack
Swigert. Extremely loud applause as Apollo 13 on
main chutes comes through loud and clear on the
television display here."

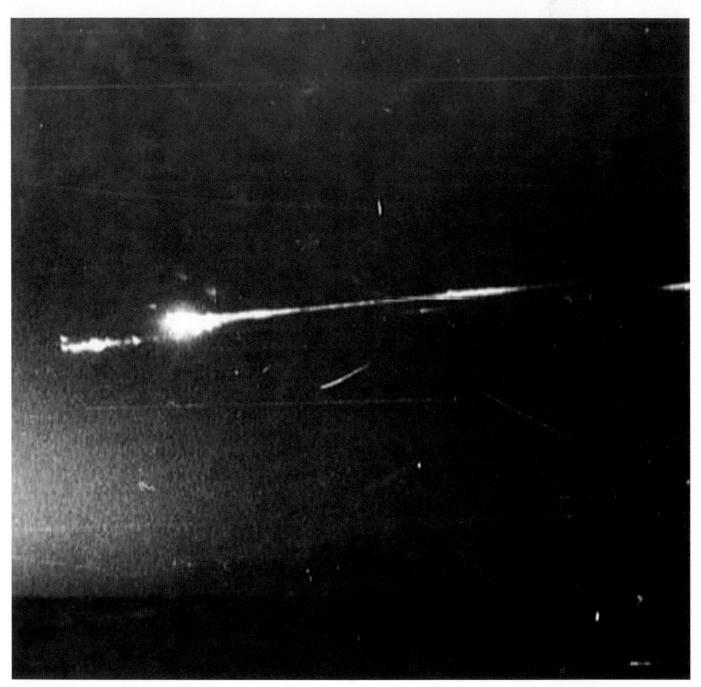

The re-entry trail of the Apollo 13 Command Module, *Odyssey*.

Stormy skies greeted the return of Apollo 13. Their pyrotechnic parachute charges worked well, despite being soaked in the cold of space for days.

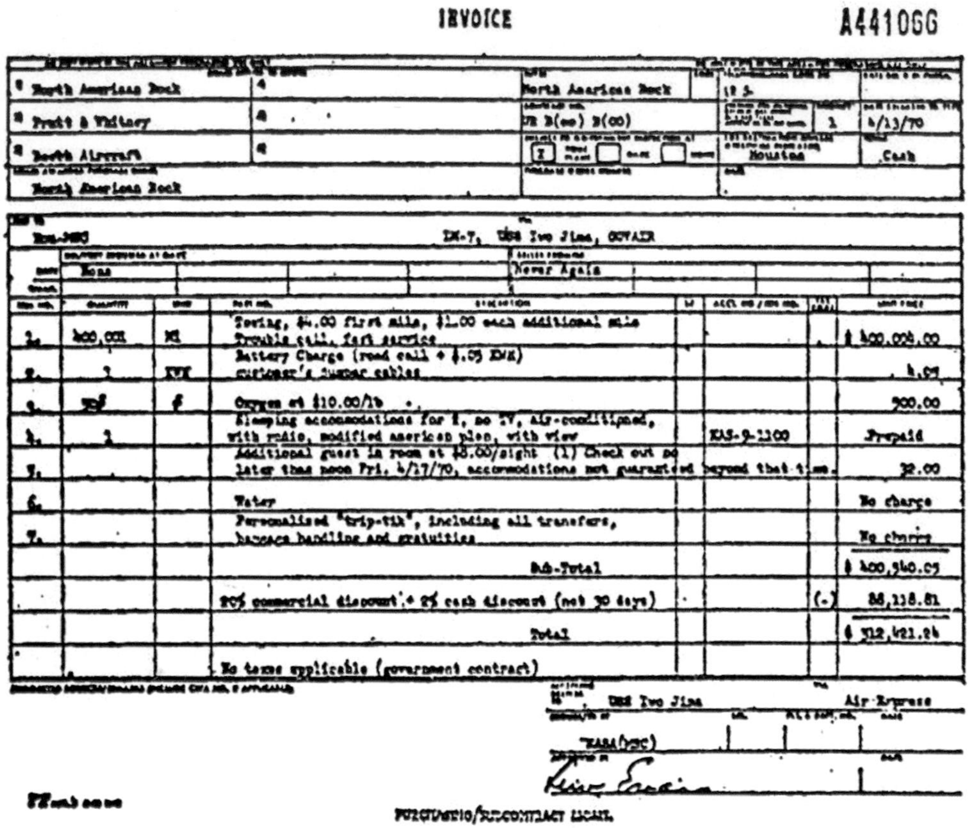

INVOICE   A441066

(Invoice form — largely illegible)

	North American Rock			North American Rock		13	
	Pratt & Whitney			JR 2(00) 3(00)			2	4/13/70
	North Aircraft					Houston		Cash
North American Rock								

EM-JEO   IM-7, USS Ivo Jima, OCVAIR

| | Ross | | | Never Again | | | |

item no.	quantity	unit	part no.	description		ACCT. no. / etc. no.		unit price
1.	400,001	MI		Towing, $4.00 first mile, $1.00 each additional mile				$ 400,008.00
				Trouble call, fast service				
2.	1	XVX		Battery Charge (road call + $.05 EXX) customer's jumper cables				4.05
3.	500	$		Oxygen at $10.00/lb				500.00
4.	1			Sleeping accommodations for 2, no TV, air-conditioned, with radio, modified american plan, with view		KAS-9-1300		Prepaid
5.				Additional guest in room at $8.00/night (1) Check out no later than noon Fri. 4/17/70, accommodations not guaranteed beyond that time.				32.00
6.				Water				No charge
7.				Personalized "trip-tik", including all transfers, baggage handling and gratuities				No charge
				Sub-Total				$ 400,540.05
				20% commercial discount + 2% cash discount (net 30 days)				(-) 88,118.81
				Total				$ 312,421.24
				No taxes applicable (government contract)				

USS Ivo Jima   Air Express

NASA(DEC)

PURCHASING/SUBCONTRACT MGMT.

Left: The infamous Grumman towing bill, which was sent as a joke to North American Rockwell for bringing their Command Module back to Earth.

Joy filled Mission Control. There were still jobs to do, but Apollo 13 and its crew were safely back home. Little flags and cigars sprung up all over the room, and for a few moments, it was delightful pandemonium. They had saved the crew.

Within an hour, Jim Lovell, Fred Haise and Jack Swigert were on the aircraft carrier *Iwo Jima* and headed for home.

Lovell subsequently retired from NASA, Haise went on to fly the Space Shuttle and Swigert prepared for a career in politics. He died of bone cancer in 1982.

Back at Grumman Aerospace, where the Lunar Modules were constructed, a staff pilot relieved the stress of the mission by preparing a practical joke. He drew up a bill and sent it to North American Rockwell in Houston, makers of the Command and Service Modules. It was a bill for towing charges, fully itemized and totalling $312,421.24.

Not an unreasonable amount when you consider the cargo.

Above: The Apollo 13 astronauts are greeted by President Nixon, who had his assistants prepare two speeches — one to use if the crew of Apollo 13 got home alive, another if they did not.

05 APOLLO 14 – EXPLORING ON FOOT

January 1971: Kennedy Space Center

After the near-disaster of Apollo 13, NASA investigated the incident until they were confident that they had discovered what went wrong. For the next mission they needed someone who could guarantee them success. Then, suddenly, Alan Shepard, who in May 1961 had become America's first astronaut in space on board Mercury 3, was back on the scene. Despite having only 15 minutes' total spaceflight experience, it was decided that he would take the reins of Apollo 14.

"Ménière's disease? What the hell is that?"

The question reverberated down the hallway of the infant Manned Spaceflight Center in Houston. The King of spaceflight, Alan Shepard himself, was grounded. The first of the Mercury astronauts would not fly again. And it was all because of a tiny imperfection of the inner ear.

Following the diagnosis in 1963, Shepard ran the Astronaut Office for NASA for eight years. But that didn't mean he had to like it. His dreams were still of spaceflight, so when a new medical procedure promised relief from the condition that had grounded

Liftoff for Apollo 14: the mission's target was the Fra Mauro highlands, which had been the objective of the doomed voyage of Apollo 13.

him, he tried it. It worked – and once again, the word echoed through the tiny spaceflight community. But this time, it was a different message about the King. This time, he went from flying a desk to winning himself a seat on Apollo 14.

On January 31, 1971, the roar of another departing Saturn V rumbled through the Florida swamps, and Alan Shepard, along with his crew mates Stuart Roosa and Edgar Mitchell, headed for the Fra Mauro highlands on the Moon.

NASA was betting on the crew of Apollo 14 to return America to space after the near-disaster of Apollo 13. Ironically, between them, their only spaceflight experience was Shepard's sub-orbital flight

Crew: Apollo 14

Alan Shepard was born in New Hampshire in 1923. He received a Bachelor of Science degree from the United States Naval Academy in 1944, and graduated from Naval Test Pilot School in 1951 and the Naval War College in 1957. He was the oldest astronaut on NASA's list and perhaps the most competitive as well.

His Lunar Module pilot, Edgar "Ed" Mitchell, was born in Texas in 1930. He earned a Bachelor of Science degree in Industrial Management from the Carnegie Institute of Technology in 1952, a Bachelor of Science degree in Aeronautical Engineering from the US Naval Postgraduate School in 1961, and a Doctorate of Science degree in Aeronautics/Astronautics from MIT in 1964. Mitchell flew for the Navy until entering the ranks of NASA astronauts in 1966.

Stuart Roosa, the CM pilot, was born in Colorado in 1933. He graduated with a Bachelor of Science degree in Aeronautical Engineering from the University of Colorado and was a test pilot at Edwards Air Force Base until joining NASA in 1966.

The Apollo 14 mission portrait. From the left: Stuart Roosa, Alan Shepard and Edgar Mitchell.

on board Mercury 3. But it didn't matter as Shepard – with more than a little help from his crew mates – would make the flight look easy.

Once in orbit around the Moon, Shepard was able to relax a bit. The crew had at least passed the hurdles that defeated the Apollo 13 landing, and would soon explore the lunar real estate that had escaped Lovell and Haise. This mission was to be a departure from Apollo 11 and Apollo 12, which had been experimental. Now it was time to get down to business. Shepard's mission was to cover more ground and undertake more exploration than both of the previous flights combined.

As the dust settled from the landing of the LM *Antares*, a quick check put them about 90 feet (27 meters) from the target landing zone. Not bad for an old man, as the CapCom later chided. Within three hours Shepard and Mitchell had completed their chores in the LM, eaten and were ready to go and "play in the snow" as Shepard put it.

Once down the ladder, the men collected a contingency sample of rock and took a better look at the terrain that surrounded them.

114:11:32 **Shepard:** "It looks as though we've landed in a fairly rough place."
114:11:36 **Mitchell:** "Yes; indeed it does. Evidenced by the fact that you dug your front landing gear into a hole..."

The LM rested on a slope, a position that did not threaten the mission, although it haunted the astronauts later as they tried to sleep. Shepard tended to roll into Mitchell's side of the tiny spacecraft cabin and wake with a start, thinking that the spacecraft was tipping over.

The two astronauts began to prepare for their first exploration of the landing zone. This trip was an intermediate step in many ways; they did not yet possess the Lunar Rover to drive in, nor the "J" series LM, which permitted much longer stays on the Moon. But they did have the Modular Equipment Transporter (MET), a wheeled device that looked somewhat like a steel-tubed rickshaw. It was a mobile tool carrier and allowed the men longer trips by foot than any previous mission. As they began to unfold the MET, little cloth patches fell out, left there by a mischievous member of the backup crew.

115:03:24 **Mitchell:** "Well, let's see... We've had visitors again."
115:03:28 **Shepard:** "Yeah. Hardly worth mentioning."
115:03:33 **Mitchell:** "Agree."

The patches had been stowed all over the LM and its gear. They each had a little roadrunner on them – a parody of the real mission patch – and had been placed there to irritate Mitchell and especially Shepard. One of the backup crew members was Gene Cernan, who had decided that Shepard was someone he could kid with mercilessly. The "hardly worth mentioning..." line was Shepard's way of fighting back.

Mitchell later recalled:

"The inside jokes from the CapComs and the little surprises that the backup crew put in the equipment; those were always important... there's always a little chuckle that comes along; there's always a little bit of levity and you can tell it."

115:03:39 **Shepard:** "Okay. Houston, as you can see, the MET is deployed properly."
115:03:45 **CapCom:** "Roger."
115:03:49 **Shepard:** "Looks like it's in good shape."

Before they could depart they also had to set up the ALSEP package, as had Apollo 12 before them. This time the radioactive fuel element came out of its protective cask just fine, and they proceeded to deploy the device with minimal difficulty.

Before long Shepard and Mitchell were back inside the LM. The next day was critical, as the two astronauts were to undertake geological traverses and other experiments for the first time. The men set up their bedding, such as it was, and attempted to get some sleep, although they were not very successful. Eleven hours later they were ready to make their epic journey toward Cone Crater, which was one of the principal objectives of their mission.

The Search for Cone Crater

Fra Mauro was not just any other Apollo destination. The Apollo 11 landing site had been chosen with

safety in mind and Apollo 12 had landed near Surveyor III. The region that Mitchell and Shepard were traversing was a part of the lunar highlands or *terrae*. The previous two landings had been in the flat *mare* areas or "seas," which were the result of comparatively recent volcanic flows, whereas the highlands were at least a billion years older. NASA believed that treasure of truly primordial rocks should be there for the taking by the crew of Apollo 14.

131:43:35 **Shepard:** "Our [line of] sight... is directly toward the center of [Cone] Crater... "
131:43:48 **Mitchell:** "Yeah, that's right over that way."
131:43:51 **Shepard:** "And it's... about 350 meters, a thousand feet."
131:43:59 **Mitchell:** "Okay. We'll start off that direction and take a look around."

The two men set off, pulling the cart-like MET behind them and full of expectation for the great unknown that lay ahead. As the terrain began to become steeper, the MET had trouble navigating the surface. Rocks and small craters covered their path, and the MET only moved over them with a lot of tugging. But neither of these driven men let these obstacles stand in their way.

131:50:21 **Shepard:** "There seem to be quite a few large rocks as we progress along here. I see rocks of up to two or three feet in size, and one would fairly easily postulate these came directly from Cone Crater. Of course, we'll get samples of these a little further along."

The Apollo 14 ALSEP package. The finned device in the foreground is the plutonium-fueled Radioisotope Thermoelectric Generator (RTG) unit for the station, which was always handled with care.

Opposite: The infamous Modular Equipment Transporter (MET). Although it was well designed, it was difficult to handle over the unexpectedly rough terrain of the Moon.

Left: A stop to gather geological samples during Apollo 14's second EVA. The band on the sleeve indicates that this is Al Shepard. After the Apollo 12 mission, all commanders sported a red band on their helmet, arms and legs to help controllers identify who they were looking at.

In a few minutes they took their first rock sample of the day. But their primary objective was to get some samples near, or on, the crest of Cone Crater, the 1,100-foot (335-meter) wide depression that appeared to be straight ahead.

132:28:47 **Shepard:** "Okay, Houston. I'm looking for a contact [recognizable feature] somewhere in here, but it's not apparent at this point. Surface texture seems to be very much the same; from the standpoint of soil-bearing properties, it's still about the same softness."

The astronauts were having trouble with navigation. Using orbital maps to navigate on the ground was often not effective, and without terrestrial cues such as haze or trees, distance was also hard to gauge. Shepard had the map in his hand.

132:31:33 **Shepard:** "Ah – that will make us right here, huh?"
132:31:50 **Mitchell:** "Well, let's take a look."

Mitchell later commented:
"We were really having trouble, on that terrain, figuring out where the heck we were. We knew where we were within a hundred meters... it was frustrating. We wasted time. And that continued. That's what slowed us down the whole rest of this thing, trying to be a little more precise about where the heck we were."

As it turned out, the two men were only about a third of the way to Cone Crater, although they believed they were halfway there. This threw off all further calculations of their position. To make matters worse, they were starting to become weary.

Shepard recalled that navigation was a challenge:
"Until we [the Apollo Program] really get a feel for navigation on the [lunar] surface, there should be some strong checkpoints to follow. First of all, it gives you a feeling of security to know where you are. You know where you are distance-wise and

The rough lunar terrain that the astronauts of Apollo 14 encountered during their ascent toward Cone Crater on the second EVA is evident in this image.

what you have left to cover. Second, there's no question in my mind that it's easy to misjudge distances, not only high above the surface... but also distances along the surface. It's crystal clear up there... it just looks a lot closer than it is."

132:33:10 Mitchell: "Okay. I think we're very close to it. I think this crater we just went by is probably it, but it's very hard to tell... I don't see anything else that might be it, unless it's the next crater up."

They set out again, compounding their errors. The geologists in the back room at Houston had no idea exactly where the two astronauts were on the map.

132:39:59 Mitchell: "Houston, as we go across here, this ground is – Al's probably previously described it – but it's very undulating. I would suspect that there is not 10 yards at the most between what were once old craters."

The going was rough. Shepard and Mitchell were also using more oxygen than planned, due to the difficult climb. As always, time was against them. Their bulky, stiff suits didn't help matters either. Later suits added some mobility, particularly a joint at the waist that allowed much more freedom of movement. But these suits were extremely hard to work in, especially with all the extra layers added for lunar EVA protection. The incline they were walking up increased to almost 20 degrees.

133:02:28 **Mitchell:** "Doggone it, you can sure be deceived by slopes here."
133:02:30 **Shepard:** "Yep."
133:02:31 **Mitchell:** "Okay, let me pull a while. You ready to go?"
133:02:32 **Shepard:** "Yeah. All set."

Mitchell took a turn pulling the recalcitrant MET cart.

133:02:53 **Shepard:** "I guess right straight up is the best way to go."
133:02:57 **Mitchell:** "Yeah, I think so."
133:03:01 **Shepard:** "Stay away from the rocks."

The two men considered the alien scenery surrounding them.

133:04:20 **Mitchell:** "Wait a minute. Yes, it is. The rim's right here. That's... that's the east [ridge]... [a] little shoulder running down from the cone... we'll have to move on around of it. This looks like easy going right here. See, there's the boulder field that shows in the photograph; it's right up ahead of us."

They continued on, but the rim of Cone Crater eluded them. Shepard was starting to have doubts about the wisdom of continuing.

133:06:10 **Shepard:** "I would say we'd probably do better to go up to some of those boulders there; document that [and] use that as the turnaround point."

The explorer in Shepard was giving way to the mission commander, and he was growing concerned.

133:09:15 **Shepard:** "I don't think we'll have time to go up there [to Cone Crater]."
133:09:16 **Mitchell:** "Oh, let's give it a whirl. Gee whizz. We can't stop without looking into Cone Crater... "
133:09:28 **Shepard:** "I think we'll waste an awful lot of time traveling and not much documenting."
133:09:33 **Mitchell:** "Well, the information we're going to find, I think, is going to be right on top."

The oldest rocks of all should have been those ejected from the bottom of Cone Crater and would reside on the ridge ahead.

133:10:10 **Mitchell:** "... how far behind our timeline are we?"
133:10:17 **CapCom:** "As best I can tell... about 25 minutes down now."
133:10:32 **Mitchell:** "Okay."
133:10:33 **Shepard:** "We'll be an hour down by the time we get to the top of that thing."
133:10:42 **Mitchell:** [Insistent] "Well, I think we're going to find what we're looking for up there."
133:10:51 **CapCom:** "Okay, Al and Ed. In view of your assay of where your location is and how long it's going

Looking back toward the LM *Antares*. Tracks from the MET are clearly visible.

to take to get to Cone, the word from the Backroom is they'd like you to consider where you are the edge of Cone Crater."

The Backroom was an area across the hall from Mission Control, where specialists, such as the geology team, dispensed advice, whether it was wanted or not, to the controllers. In this case, Mitchell thought they were giving up too easily.

133:11:13 **Mitchell:** "I think you're finks!"
133:23:23 **CapCom:** "And, Ed and Al, we've already eaten in our 30-minute extension and we're past that now. I think we'd better proceed with the sampling and

continue with the EVA."
133:23:37 **Mitchell:** "Okay."
133:23:40 **Shepard:** "Okay. We'll start with a pan from here. I'll take that."
133:23:47 **Mitchell:** "All right, I'll start sampling."

Mitchell later recalled:
"It was terribly, terribly frustrating; coming up over that ridge that we were going up, and thinking, finally, that was it; and it wasn't – suddenly recognizing that, really, you just don't know where the hell you are. You know you're close. You can't be very far away. You know you got to quit and go back."

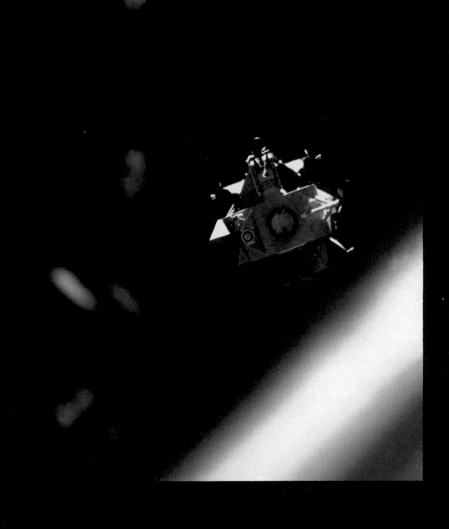

Apollo 14's LM *Antares* as seen from the CM *Kittyhawk*.

Shepard also remembers:

"We were probably within 150 to 300 feet – depending on these two boulder fields – of the rim and still were not able to see it. That was a pretty good-sized lunar feature, and to be that close to the top of the thing and not see it. That is just part of the navigation problem."

Mitchell added:

"At this point, in spite of personal frustration – and I know Al felt frustrated in the same way – to have us stop at that point and turn around and come back was the proper decision."

The two trudged back to the LM *Antares*, still pulling the MET. This was the last Apollo mission that would have to explore on foot, as the Boeing-built Lunar Rover would allow future moonwalkers to explore further and more completely than before.

Once they returned to Earth and the mission was reviewed, the errors were deduced and a stunning conclusion was reached. On the way back from their search for the summit of Cone, they had unknowingly passed within 65 feet (20 meters) of the rim. As one of the participating geologists later put it to Shepard, "You weren't lost, and you didn't know it."

Before departing the Moon, Shepard had one last trick up his sleeve. To the delight of duffers worldwide, he attached a six-iron to the sample collector handle and whacked a golf ball as far as it would go… as Shepard himself observed: "Miles and miles and miles."

The last of the short-stay "H" lunar missions was complete, and NASA was back on track. After splashdown, the three astronauts waited out their quarantine in their special trailer. As irritating as this could be, Apollo 14 was the last time returning astronauts had time to collect their thoughts before being besieged by the media, as NASA decided that the risk of contamination by the moonwalkers was about zero.

In the trailer, the crew reflected on the voyage. Stu Roosa later told North American Rockwell that their improvements on the Command Module had "saved the Apollo Program." Ed Mitchell had experienced a revelation during the flight and later set up an organization called The Institute of Noetic Sciences to study human consciousness. And Al Shepard was, at last, satisfied with his NASA career and left the organization to go into a variety of business ventures.

It was the eve of NASA's greatest triumph – the Apollo "J" missions. The King of spaceflight, with the assistance of two rookies, had given the Apollo Program wings again.

04 APOLLO 15 – THE GENESIS ROCK

July 1971: Hadley Rille

The Apollo 14 mission had put NASA back on track, but the geological puzzle remained regarding the formation of the Moon. This mission – the first with a Lunar Rover – would continue the quest to discover the secrets that lay beneath the lunar dust.

The next three Apollo missions ended the program on a high note. These were the "J" series missions, which allowed much longer stays on the Moon. The Lunar Module had been modified and had more life support capacity, and the moonsuits had been altered to allow more mobility and flexibility. But there was another big difference – these missions had a car.

When Boeing Aircraft received the contract for the Lunar Rover in 1969, it gave them scarcely two years to design and build the most complex wheeled vehicle in history. It had to be light, robust and ultra-dependable. It also had to come in on-budget and

The Lunar Roving Vehicle (LRV) gets a once-over from engineers. The vehicle – a $20 million project – was completed ahead of schedule by Boeing and General Motors.

work perfectly. Boeing immediately teamed up with the obvious choice for the venture, General Motors, and the marvel of automotive engineering was completed months ahead of schedule. The Rover underwent an immediate program of hard testing and, although some modifications were required, it was probably the most reliable Chevy in history.

The $20 million contract resulted in a vehicle that was light, compact and foldable. It could carry two times its weight at a maximum of 10 miles (16 kilo-meters) per hour for up to 60 miles (97 kilometers), cross a 28-inch (70-centimeter) crevasse and climb a 25° slope. The batteries worked for up to 78 hours and its operational temperature ranged from -200° Fahrenheit (-93° Celsius) to more than +200° Fahrenheit (93° Celsius).

Crew: Apollo 15

Riding the rocket at the launch of Apollo 15 on July 26, 1971 was an all-star crew. With his rakish good looks and model physique, Commander Dave Scott was described as an all-American boy. He was demanding and a perfectionist and he intended this to be the best Apollo mission yet. Scott received a Bachelor of Science degree from the United States Military Academy and two Master of Science degrees from MIT. He graduated from the Air Force Experimental Test Pilots School and joined NASA in 1963. Prior to this flight, his most noted NASA moment was alongside Neil Armstrong on Gemini VIII, during which the two astronauts barely survived an out-of-control spacecraft and emergency re-entry.

The Lunar Module pilot was Jim Irwin. A devoutly religious individual, Irwin was born in Pennsylvania in 1930. He received a Bachelor of Science degree in Naval Science from the United States Naval Academy in 1951 and a Master of Science in Aeronautical Engineering from the University of Michigan in 1957. He joined the Air Force upon graduation from the Naval Academy and migrated to NASA in 1966.

Destined to remain in the Command Module *Endeavour* was Al Worden. Born in Michigan in 1932, he held a Bachelor of Military Science degree from the United States Military Academy at West Point and a Master of Science in Astronautical and Aeronautical Engineering from the University of Michigan in 1963. He graduated from the United States Military Academy in June 1955 and served as an instructor at the Aerospace Research Pilots School. He joined NASA in 1966.

The crew of Apollo 15. From the left: Dave Scott, Al Worden and Jim Irwin; the Lunar Rover can be seen to the left of the crew.

It was truly an all-weather, all-terrain vehicle for the Moon. It even had a satellite radio system, a device unheard of in 1969. Its target satellite was the Earth, 240,000 miles (386,232 kilometers) away.

As the crew of Apollo 15, David R. Scott, Alfred M. Worden and James B. Irwin, prepared for departure, their Lunar Roving Vehicle (LRV) – which the crews simply called "the Rover" – was squeezed into the Lunar Module *Falcon*. The new LM would be truly at home on the Moon, allowing three moonwalks and stays of up to three days.

Four days after their launch, Scott and Irwin were gliding over the lunar Apennine Mountains in *Falcon*. The LM's tiny navigation computer fired the thrusters to twist the craft to vertical, and the two men finally got a view of the site where they would land within minutes. Ahead of them was a vast plain, with a valley cut into the middle of it – Hadley Rille. After a nominal descent, *Falcon*, which was heavier than its predecessors, set down with a notable "wham" into the lunar dust.

104:42:14 **Irwin:** "15 at one. Minus one, minus one; six per cent fuel."
104:42:22 **Irwin:** "10 feet. Minus one."
104:42:27 **Irwin:** "Eight feet. Minus one."
104:42:29 **Irwin:** "Contact... Bam!"

Irwin later noted:
"We did hit harder than any of the other flights! And I was startled, obviously, when I said, 'Bam!' [Laughing] And I think Dave didn't particularly appreciate my comment that he made a hard landing on the Moon!"

Jim Irwin with the Rover, which is parked at the edge of Hadley Rille.

Dave Scott, Commander of the mission, salutes the deployed flag during the first EVA.

104:42:36 Scott: "Okay, Houston. The *Falcon* is on the plain at Hadley."
104:42:40 CapCom: "Roger, *Falcon*."
104:42:48 Irwin: "No denying that. We had contact!"

Two hours later Scott tried an exercise that had been devised by the program's geologists. They wanted a high-viewpoint survey, so he opened the hatch on the top of the LM – normally used only for docking with the Command Module – and stood up for a 360° view.

106:48:51 Scott: "Oh boy, what a view…"After half-an-hour of photography and landmark identification, Scott came back inside and closed the hatch.

He later recalled:
"I might say, in conclusion, that the SEVA [Standup EVA] was a very useful thing... at this time, I wasn't sure where we were located. Although I could see prominent features, I was relying on the Sun compass to give us the data for triangulation to spot our point, because there was nothing in the immediate vicinity which was recognizable."

The two men tidied up the LM, ate and prepared to sleep. On the Moon, their bodies weighed less than 30 pounds (14 kilograms).

109:57:42 Scott: "Houston, Hadley Base… we're all tucked into the hammocks, and we'll see you in the morning."
109:58:19 CapCom: "Roger, Dave. Good night and don't fall out."
109:58:26 Scott: "No way, Bob, no way. There's no place to go if I did!"

Scott later reflected:
"A lot of people said, 'Gosh, how can you go to the Moon and go to sleep, without getting out?' Well, Jim and I used to talk about some of these quasi-philosophical things, so we were on the same frequency. Neither one of us thought we'd have any problem. I mean, it is a big deal, but it's also your job. And, if you're tired, you go to sleep."

Irwin weighed in:
"The first night's sleep on the LM was the best night's sleep I had on the total flight."

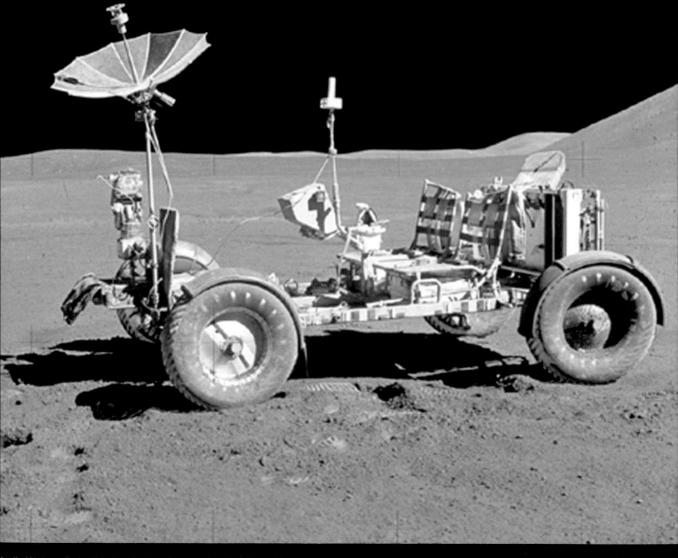

Apollo 15's Lunar Roving Vehicle just prior to the exploration of the Hadley Rille formations.

The First EVA: The Rover

The next morning, Scott and Irwin rose early, went outside and deployed the Rover. Then they were off and running. By 10 a.m., the first Lunar Road Rally was underway.

121:45:02 **Scott:** "Whew! Hang on."
121:45:06 **Irwin:** "And we're coming around left... heading directly south right now to miss some craters off to our right – very subdued craters."
121:45:24 **Scott:** "Okay, I'm going to take a little zigzag here..."

Although he drove at an average of about eight miles (13 kilometers) per hour, Scott had to attempt some pretty fancy maneuvers to cross the "plains" on which they had landed. The surface was much rougher than expected, and even at these low speeds and at one-sixth gravity, driving the Rover was an adventure.

121:51:34 **Scott:** "Okay. Whoa. Hang on."
121:51:42 **Irwin:** "Bucking bronco."
121:51:43 **Scott:** "Yeah, man. You back off on the power, it keeps right on going."
121:53:50 **Irwin:** "Boy, it really bounces, doesn't it?"
121:53:54 **Scott:** "Well, I think there's sort of a... the rear end breaks out at about 10 to 12 clicks."

The Rover was fun to drive. It also had an onboard navigation system, to eliminate the navigational difficulties that had bedeviled Apollo 14. Hadley Rille, the valley that cut across their landing zone like a giant scar, lay ahead of Scott and Irwin.

121:57:21 **Irwin:** "Hey, you can see the rille! There's the rille."
121:57:23 **Irwin:** "We're looking down and across the rille, we can see craters on the far side... "

They continued the drive, as Irwin marvelled at the grandeur before him.

122:04:18 **Irwin:** "I can see the bottom of the rille. It's very smooth. I see two very large boulders that are right on the surface, there."
122:06:33 **Scott:** "I might add to Jim's comment, that the near side of the rille wall is smooth without any outcrops, there by St George [crater], and the far side has got all sorts of debris. It almost looks like we could drive down in on this side, doesn't it?"

The CapCom, Joe Allen, was probably not really concerned that they would, but just to be certain...

122:06:48 **CapCom:** "Uh... stand by on that, Dave."
122:06:49 **Irwin:** "I'm sure we could drive down; I don't think we could drive back out."

Both men laughed. Scott and Irwin were having a bit of fun teasing the flight controllers. Although it had never been a part of the final plans to drive into the rille, it had been discussed. But to attempt to do so at this point would have resulted in dozens of heart attacks back at Houston. After taking some samples they headed back to the *Falcon*, where they began work on the ALSEP deployment and, after a frustrating period of drilling through the Moon's hard surface, went inside for the "night."

The Second EVA: The Mountains of the Moon

The next day the two astronauts headed toward the base of Mount Hadley, which rose 14,765 feet (4,500 meters) above them. There was a noticeable incline, and when they made the first stop, the panorama was breathtaking. They continued on toward Station 6, which was about 400 feet (122 meters) above the valley floor.

144:59:46 **Irwin:** "Bearing 345... boy, you know, looking upslope, look how much more hummocky it is. It's just a different terrain."
145:00:13 **Scott:** "It sure is. It sure is. Pretty hummocky and driving is much sportier."
145:00:25 **Irwin:** "Yep."
145:00:27 **Scott:** "Ooh. Hang on. Hang on; there we go. This Rover is super!"

The two had reached the boulder that was their destination. They were now on a slope, and Irwin wanted to make sure they stayed above the rock – presumably, in case it rolled downhill. They prepared their sampling tools as they headed toward the boulder. The hike was more treacherous than either had planned.

145:04:19 **Scott:** "Okay; let's attack that boulder. You got your hammer?"

Irwin was looking at the steep slope behind them, leading back to the Rover.

145:04:23 **Irwin:** "Gonna be a bear to get back up here, you know."

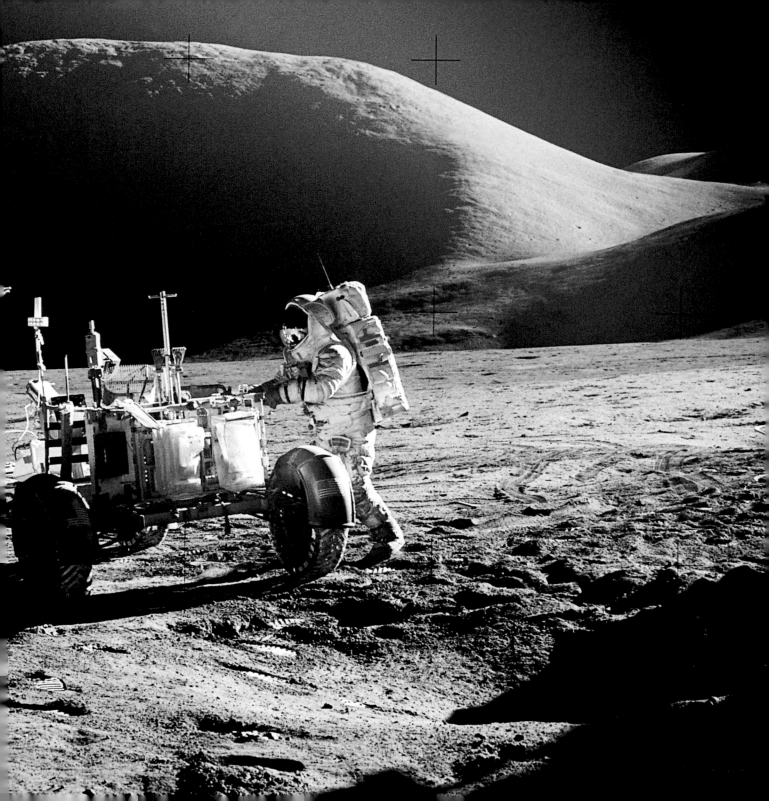

Dave Scott collects a geological sample not far from a crater named Elbow.

In Houston, the controllers were getting nervous. They knew they had an overachiever in Scott, and didn't want any accidents at this stage.

145:04:29 **CapCom:** "Hey, troops, I'm not sure you should go downslope very far, if at all, from the Rover."
145:04:37 **Scott:** "No, it's not far. Let me try it, Jim, you just stay there."

Irwin sounded doubtful.

145:04:40 **Irwin:** "I *think* we can sidestep back up."
145:04:43 **Scott:** "It's not that hard."

The CapCom, who could talk to them but was not receiving a TV image, was not so confident.

145:05:24 **CapCom:** "Okay, Dave. How's the footing?"
145:05:26 **Scott:** "Well, the footing is all right, except that you have to work pretty hard to get back up."

Mission Control was tense and quiet. All eyes were on the blank screen. Up on the Moon, two astronauts were risking it all for this rock – and everyone knew that one false move in an Apollo-era moonsuit could be your last.

145:11:31 **Scott:** "Okay, Jim, you can come on down now."
145:11:33 **Irwin:** "Yeah. [Pause] I'd estimate a, what, 20° slope?"
145:11:45 **Scott:** "Closer to 15, probably."

145:11:58 **Irwin:** "See, the back wheel's off the ground."

The Rover was parked on such a steep incline that it was beginning to tip.

145:12:02 **Scott:** "Yeah. I think I'll get back on. Tell you what, Jim. We'd better abandon this one."
145:12:14 **Irwin:** "Afraid we might...lose the Rover?"
145:12:42 **Irwin:** "Oh, you really... Let me hold that Rover and you come up and look at this, because this rock has got green in it, a light green..."

The Rover had lost traction and was starting to slide downhill. Scott held on to it for dear life, while Irwin continued to take samples. Houston advised temperance.

145:13:34 **CapCom:** "Dave and Jim, use your best judgment here, the block's not all that important..."

The astronauts took their sample and got back into the Rover to depart. It was a tricky outing, but during the adventure they had retrieved a piece of green rock, something that nobody had expected to find on the Moon. However, the best was yet to come.

The Genesis Rock

Driving toward Spur Crater, Irwin noticed something unusual. A small rock was perched almost perfectly on top of a boulder – it was as if it had been placed there by someone and was just waiting to be found.

Left: Sample number 15415, dubbed "the Genesis Rock." It was a piece of very old anorthosite, some of the original material from the lunar crust, dated at 4.5 billion years old.

Opposite: Apollo 15 at the moment of splashdown: the capsule hit the ocean harder than most Apollo flights, as one of the parachutes did not open completely, however nobody was hurt.

145:42:41 **Irwin:** "Oh, man!"

When they first sighted it, the rock was gray...

145:42:41 **Scott:** "Oh, boy!"

As they dusted the rock, it became white...

145:42:42 **Irwin:** "I got..."

And a crystalline structure revealed itself.

145:42:42 **Scott:** "Look at that!"

It was what they had hoped to find.

145:42:44 **Irwin:** "Look at the glint!"
145:42:45 **Scott:** "Aaah."

Irwin was unable to talk and laughed out loud. He was electrified.

145:42:47 **Scott:** Guess what we just found. Guess what we just found! I think we found what we came for."

145:42:53 **CapCom:** "Crystalline rock, huh?"
145:42:55 **Scott:** "Yes, sir. You better believe it... make this bag, 196, a special bag."

And special it was. The rock was a chunk of anorthosite, which formed part of the primordial crust of the Moon, and its discovery was a major aim of the Apollo 15 mission. When the geologists back on Earth tested the rock, it dated back 4.5 billion years, only about 100 million years younger than the origins of the Solar System itself. Its official name was Sample number 15415, but it quickly became known as "the Genesis Rock."

Soon after the discovery Scott and Irwin returned to the LM and blasted off from the Moon. The two men were ecstatic, and Worden had to listen to their excited chatter all the way back to Earth. He didn't mind one bit.

Subsequent flights had other unique finds, and slightly older rocks were collected, the oldest of which was discovered by Apollo 17. However, none of those later moments were quite so perfect as this. Perhaps it was because it was the first such rock discovered – or perhaps it was, as Jim Irwin thought, the way it was found – but somehow, it seemed almost like divine intervention.

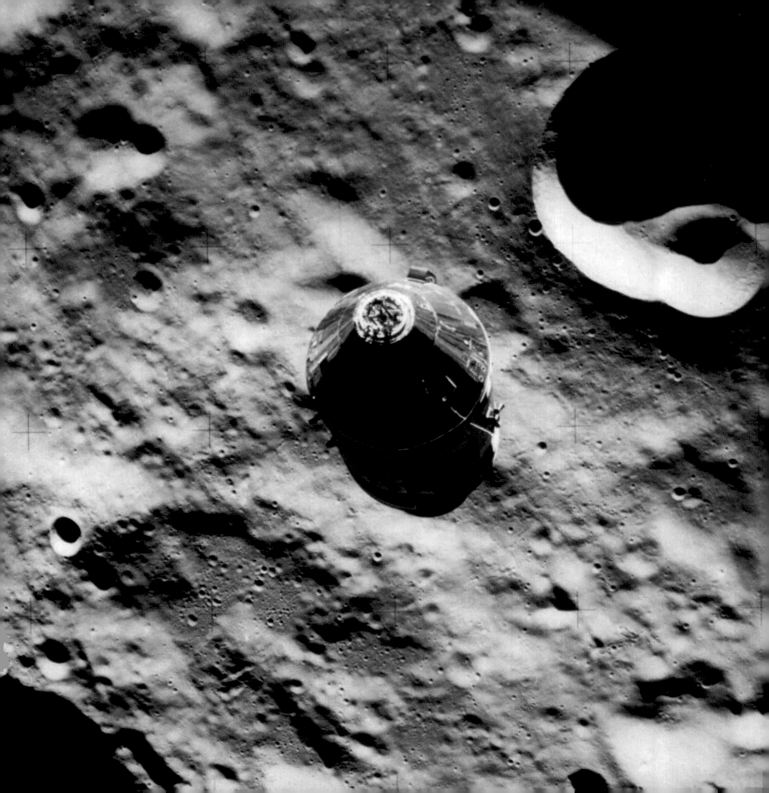

03 APOLLO 16 – THE LUNAR GRAND PRIX

April 1972: The Descartes Highlands

For the second time in the space of a few months, a Lunar Rover tore up luna firma. It was the beginning of another all-too-brief period of intense exploration on the Moon – the Apollo 16 mission, which had launched on 16 April, 1972. This time, Charlie Duke was in the Rover's passenger seat and John Young was at the wheel.

123:02:36 **Duke:** "How's it driving, John? Pretty easy?"
123:02:38 **Young:** "Darn good."
123:02:40 **Duke:** "Hey, man. We could just go, babe. I'm really cinched into this moose."

It was the beginning of their first EVA. The "moose," as Duke had called the Rover, was performing flawlessly. As his partner, John Young, was driving, Duke scrambled around in his seat, taking pictures and reading maps. Overhead, Ken Mattingly circled in lunar orbit in the Command Module *Casper*.

123:04:29 **Young:** "Charlie, you hit my arm."

Duke kept working and talking to Houston, oblivious to disturbing the normally reasonable Young.

123:04:50 **Young:** [Annoyed] "Quit hitting my arm!"

Young later recalled:
"I was scared to go more than four or five kilometers an hour. Going out there, looking dead ahead, I couldn't see the craters. I could see the blocks all right and avoid them. But I couldn't see craters... maybe sometimes I got up to six or seven kph, but I ran through a couple of craters because I flat missed [seeing] them until I was on top of them..."

Apollo 16's Command Module *Casper* as seen from the Lunar Module *Orion*.

Crew: Apollo 16

John Young was born in 1930 in San Francisco. He earned a Bachelor of Science degree in Aeronautical Engineering with highest honors from Georgia Institute of Technology in 1952. After test pilot training at the US Navy Test Pilot School in 1959, he was assigned to the Naval Air Test Center for three years and was selected as one of the second astronaut intake group in 1962. He flew the first manned Gemini mission with Gus Grissom in 1965, and went on to fly Gemini X with Mike Collins in 1966 and Apollo 10 with Tom Stafford and Gene Cernan in 1969. Now he was in command of an Apollo Moon landing.

Jocular and easy-going compared to many of his fellow astronauts, Charles M. Duke, Jr was born in North Carolina in 1935. He received a Bachelor of Science degree in Naval Sciences from the US Naval Academy in 1957 and a Master of Science degree in Aeronautics from MIT in 1964. Before he became an astronaut in 1966, Duke worked at the Air Force Aerospace Research Pilot School as an instructor.

Thomas "Ken" Mattingly was the CM pilot. Scrubbed from Apollo 13 because of a suspected case of measles, this was his last chance at a lunar mission. Born in 1936 in Chicago, he received a degree from Auburn University in 1958. Mattingly was selected by NASA in 1966.

The crew of Apollo 16. From the left: Ken Mattingly, John Young and Charlie Duke.

At Mission Control, managers decide whether or not to approve the Apollo 16 landing. Ken Mattingly, orbiting the Moon in the Casper, was having engine trouble. In the end, the problems took care of themselves. Seated at center is Christopher Kraft, Jr, Director of the Manned Spacecraft Center. Standing, from left to right, are Dr Rocco Petrone, Apollo Program Director, Office of Manned Space Flight; Capt. John Holcolmb, Director of Apollo Operations; Sigurd Sjoberg, Deputy Director, MSC; Capt. Chester Lee, Apollo Mission Director; Dale D. Myers, NASA Associate Administrator for Manned Spaceflight and D. George M. Low, NASA Deputy Administrator.

"... When you got to a ridge, you couldn't tell if it was a drop-off, or whether it was a smooth, shallow ridge. In a couple of cases, you couldn't see there was a ridge. I didn't care for that much."

Due to a combination of the Sun's angle, the high reflectance of the lunar soil and the direction they were driving, Young was barely able to see any of the features on the route ahead. He was very concerned about breaking the Rover. Duke, on the other hand, seemed to be having a great time.

123:02:48 **Duke:** "This seatbelt is great. It seems to be taking it with no problem."

They bounced over some more obstacles.

123:06:57 **Duke:** "Right back there, John. Boy, it's really hard [to navigate]. There's an interesting rock. A layered [rock], really dust covered... turn left, John. There's a crater over there, a big one."
123:07:18 **Young:** "Boy, that is a biggie."

They could see a feature known as Spook Crater.

123:07:28 **Duke:** "Boy, that is a biggie. Okay, here is Spook."
123:07:36 **CapCom:** "Okay."
123:07:38 **Duke:** "And that is a biggie!"

Both men laughed like kids in the Solar System's biggest sandbox. Despite Young's concerns, they were having the time of their lives. But navigating was still a touch-and-go affair, for, although the Rover had a navigation system, the astronauts were still dependent on visual landmarks. And virtually all of the photographic maps they used had been taken from orbit. Things on the ground tended to look very different.

Young later reflected:
"You know, we never encountered any of these features on the geology map... I think that's because those photo-analysis guys were reaching for and pulling out features that weren't there. I mean, I looked for these things and sure enough, if you really imagined it, you could see something there."

Their navigation suffered because what geologists on the ground thought should or could be present on the Moon's surface wasn't necessarily evident when the astronauts actually arrived there. It was reminiscent of the Mars maps that Percival Lowell created at the beginning of the twentieth century, when he imagined canals on the red planet, a result of seeing lines of craters and landforms that resembled an engineered aqueduct. While the Apollo geologists had much more sophisticated and higher-tech tools at their disposal, they may have drawn similar conclusions about the Descartes area of the Moon.

The astronauts continued on to a sampling stop at Plum Crater.

123:11:24 **Duke:** "There it is, right there."
123:11:26 **Young:** "That's Plum?"
123:11:28 **Duke:** "Yeah!"
123:11:30 **Young:** "That ain't even on the rim."
123:11:32 **Duke:** "Well... yeah, it is. It's right... here... on top of the rim."
123:11:39 **Young:** "Okay."
123:11:40 **Duke:** "Hey, stop. It's going to be terrible walking on this thing. Why don't we go turn around and go back up on the rim where it's level?"
123:11:49 **Young:** "Suits me."

But it was the wrong place. As any moonwalker can tell you, all lunar craters tend to look alike, and with no visual cues for distance, such as trees or houses, it's very hard to tell how big things on the Moon are.

123:16:49 **Duke:** "Man, it all looks the same, doesn't it?"
123:16:52 **Young:** "Sure does."

The misidentification didn't dampen Duke's enthusiasm one bit, though. Throughout the mission, he was the most exuberant astronaut anyone could remember since Pete Conrad of Apollo 12.

123:23:38 **Young:** "Man, you can't believe this territory."
123:23:44 **Duke:** "It's all up and down. We're gonna be a little close here [to Plum], John, but that's okay."
123:23:54 **Duke:** "Okay, we're parking right on the rim of Plum."

They had found their target. Duke later recalled looking deep into Plum Crater and marveling at just how far down it went. But he had no desire to climb into it. The two picked up some representative rock samples, and Young used his hammer to whack off a piece of a boulder.

123:58:35 **Duke:** "Ah, here he comes, folks! He's got the hammer out. I knew he couldn't resist."

They both imitated a B-movie slasher preparing for the kill, cackling evilly. Duke later recalled:

"In training, every time we got to a rock he'd whack it. He was really excited about using the hammer when we got to the Moon, 'cause he'd used it a lot in training, whacking off chunks of rock."

123:59:05 **Duke:** "Hot dog! He did it!"

One of Apollo 16's mission objectives was to look specifically for volcanic rocks. The debate over whether the Moon's surface was more volcanic in origin or had formed as a result of meteorite impacts had raged for years, and geologists hoped that the Descartes Highlands might answer this question.

The geologists in the backroom leaned forward, hoping it was not another breccia, or impact-derived, rock.

124:00:13 **Young:** "Yeah, it's a breccia, Houston."
124:00:15 **Duke:** "Yeah, uh-huh."

Charlie Duke taking a sample of rock at Plum Crater.
John Young can be seen in Duke's visor taking the photo.

124:00:17 Young: "Or a welded... "

For a moment spirits lifted, it might be a welded tuff
(a piece of volcanic rock).

124:00:22: Duke: "No, that's not right. It's a breccia...
it's just a one-stage breccia."

No volcanic rocks here. Duke took some
documentation photos anyway, then noticed that
Young was standing right on the rim of the crater.

124:03:01 Duke: "John, you are just beautiful. That is
the most beautiful sight."

Young was caught off guard by the unique comment.

124:03:07 Young: "What's that?"
124:03:08 Duke: "You standing there on the rim of
that crater."

Sarcasm traversed the 240,000 miles (386,232
kilometers) distance from home.

124:03:09 CapCom: "Doggone. I've never heard John
described as beautiful."
124:03:12 Duke: "Well, he's not really... Well, actually,
he is on this thing; I'll tell you."

He referred to the rim Young had mounted. Disregarding his
own remark, he waved to Young who was taking a photo.

124:03:21 Duke: "Hi, there!"

They completed a few more stops and before long
headed back to the LM *Orion* to eat, rest and sleep.
They had two more moonwalks to do over the next
couple of days, but first, they had some road-testing
to do for NASA. Duke stood close to the LM, and
Young drove the Rover at maximum speed. Houston
wanted some film of the vehicle going through
the paces.

124:57:10 Duke: "Man, you are really bouncing!"
124:57:14 CapCom: "Is he on the ground at all?"
124:57:20 Duke: "He's got about two wheels on the
ground. There's a big rooster tail out of all four wheels.
And as he turns, he skids. The back end breaks loose
just like on snow. Come on back, John... I'll tell you,
Indy's never seen a driver like this."

They went for a second run.

124:59:03 Duke: "Man, that was all four wheels off
the ground, there. Okay. Max stop."
124:59:12 Young: "Okay. I don't want to do that."
124:59:13 Duke: "Okay. Excuse me?"
124:59:16 Young: "They say that's a no-no."

The designers had warned against slamming on the
brakes.

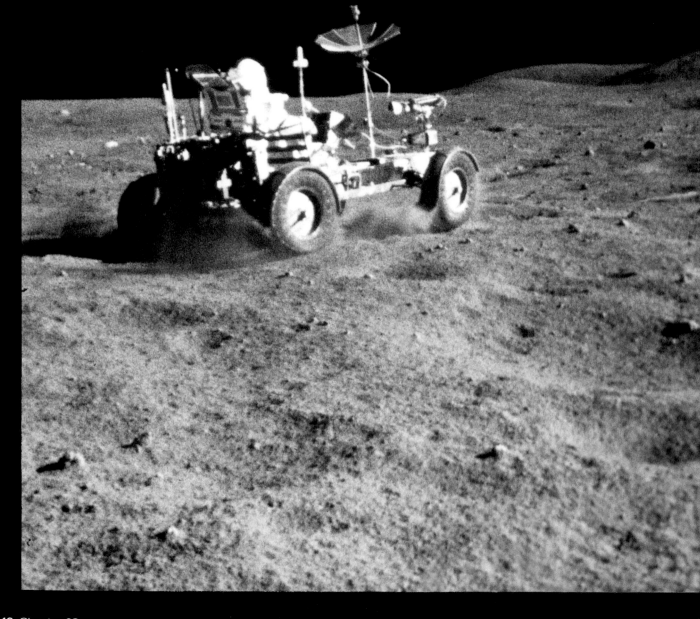

John Young taking the Lunar Rover through its paces as Charlie Duke films the experiment on 16mm film.

124:59:22 **Duke:** "Okay…"
124:59:27 **Young:** "Okay. I have a lot of confidence in the stability of this contraption."
124:59:30 **Duke:** "Me, too."

Young later related:
"I didn't get up to any great speed – maybe 10 clicks at the most – but the terrain around there was too rough and too rocky for that kind of foolishness."

Soon they were back inside the LM and readying to go to sleep for the night. Then a problem arose.

Gas

Inside the LM, Mission Control reminded them about their diet. They had been instructed to drink a lot of orange juice to reduce the possibility of experiencing heart irregularities, which had plagued the first extended mission, Apollo 15. Young was fed up with drinking the acidic beverage.

128:19:04 **Young:** "I'm going to turn into a citrus product is what I'm gonna do."
128:19:09 **CapCom:** "Oh, well; it's good for you, John."
128:19:15 **Young:** "Ever hear of acid stomach, Tony [CapCom]?"
128:19:17 **CapCom:** "Well, I don't know about that."
128:19:30 **Young:** "Okay, and I think I've got a pH factor going for about three right now."

They continued to talk as they did housekeeping, thinking that the microphone to Houston – and possibly the rest of the world – was off.

128:50:37 **Young:** "I have the farts, again. I got them again, Charlie. I don't know what the hell gives them to me… I think it's acid stomach. I really do."
128:50:44 **Duke:** "It probably is."
128:50:45 **Young:** [Laughing] "I mean, I haven't eaten this much citrus fruit in 20 years! And I'll tell you one thing, in another 12 f***ing days, I ain't never eating any more. And if they offer to [give] me potassium with my breakfast, I'm going to throw up! I like an occasional orange. Really do. But I'll be darned if I'm going to be buried in oranges."

They continued for a while, before the CapCom could break in.

128:53:58 **CapCom:** "Okay, John. We have a hot mike."

Young was startled.

128:54:07 **Young:** "Ah – how long have we had that?"
128:54:10 **CapCom:** "It's been on through the debriefing."

They reconfigured the microphone and returned to talking about business. Four hours later they were asleep.

Stone Mountain

It was a long drive but the two explorers now stood on higher ground than any other Apollo crew. They were 500 feet (153 meters) up Stone Mountain, about three miles (five kilometers) from the LM. The two men observed the scene below and took numerous photographs from their vantage point.

144:09:46 **Duke:** "Tony [CapCom], you just can't believe this! You just can't believe this view! You can see the Lunar Module, you can see North Ray with boulders on the southwest side; and where Station 12 is, there's one huge boulder that's going to be just great."
144:16:27 **Duke:** "Wow! What a place! What a view, isn't it, John?"
144:16:30 **Young:** "It's absolutely unreal!"
144:16:34 **Duke:** "We've really come up here, Tony. It's just spectacular. Gosh, I have never seen…. All I can say is 'spectacular,' and I know y'all are sick of that word, but my vocabulary is so limited."
144:16:50 **CapCom:** "We're darn near speechless down here…"

Duke and Young took samples at various points, trying to find some evidence – any evidence – of volcanic rocks. The geology team had hoped to find some in this area. Instead they had to settle for a different geological prize.

House Rock

After another night in the LM – this one was methane-free – the two explorers proceeded with their third and last EVA. Since they had arrived, they had been eyeing a huge boulder, which was situated far off and close to a formation known as North Ray Crater. It was time to get a piece of it.

167:36:10 **Young:** "Look at the size of that biggie… it is a biggie, isn't it. It may be further away than we think because… "
167:36:17 **Duke:** "No, it's not very far… "
167:36:19 **Young:** "Theoretically, huh?"
167:36:20 **Duke:** "Yeah."

They had been fooled before. It was very difficult to tell how big the boulder was – or how far away it was situated. The astronauts drove as far toward it as they could, then dismounted the Rover and headed off toward the horizon. In Mission Control, all eyes were on the TV screen. The smaller the two astronauts got, the larger the rock had to be.

167:43:11 **Duke:** "Well, Tony, that's your 'house rock' right there."
167:43:14 **CapCom:** "Very good."

Young soon joined Duke next to the 40-foot (12-meter) high rock.

167:43:52 **Duke:** Okay, John... looky here. Can we whack with the... see, look at that. See, it's glass coated, and this is just fractured off. We could pull that off. Big chunks of [it] that'll come right off."

The irreverent Duke got his sample. The 213 pounds (97 kilograms) of rocks they collected was a treasure trove of information to the geologists. And if they were unable to vet the volcanic theories so many held dear, they were able to ascertain that the region that Apollo 16 visited was covered with material from an impact in the Imbrium Basin, over 600 miles (966 kilometers) away. An event of that caliber was difficult to imagine, but the evidence was there, and showed up later under the microscope in the Lunar Receiving Lab. Geologists had to adjust their thinking toward the Moon. Or, as Ken Mattingly put it from the CM *Casper* in orbit above the Moon, "Well, back to the drawing boards, or wherever geologists go...."

Above: Sample number 68815 as seen in the Lunar Receiving Lab. It is easy to see why identification of rocks by observers on the Moon was not a simple task: they tend to look very much alike. Note the gas fissure in the center of the rock.

Left: The ascent stage of *Orion*. Damage to the back of the cabin is visible and probably stems from liftoff. This sheet metal covers electronics but did not impinge on the integrity of the cabin itself.

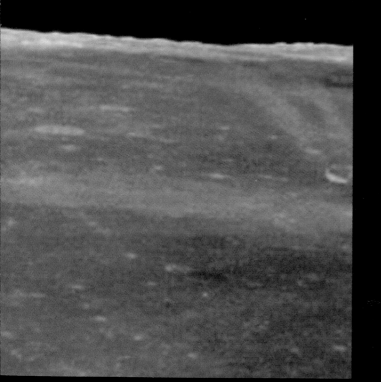

02 APOLLO 17 –
THE LONE SCIENTIST

December 1972: Taurus-Littrow

Apollo 17 was the last flight of the lunar program. Numbers 18, 19 and 20 were now just forgotten numbers on the flight manifest. As this was the final mission, NASA decided to put a true scientist on board, and Harrison H. "Jack" Schmitt, with a Ph.D in Geology, soon found himself on the lunar surface. This last mission proved to be one of the most satisfying.

They called him "Doctor Rock." Far from an insult, it was an apt description for the only scientist to visit the Moon. Of the 12 men that roamed the lunar surface during the Apollo missions, all were pilots and military men except for Jack Schmitt. With a doctorate in geology from Caltech, he was the perfect choice for Apollo 17.

On the first of three moonwalks, time flew far too quickly, and even with both Schmitt and Cernan working furiously to complete their tasks, the geologist knew they would miss something. With this in mind, he was vaguely annoyed when Cernan, as enthusiastic as

December 7, 1972 – the launch of the final lunar mission, Apollo 17. Due to a specific launch window, it was Apollo's only night launch.

anyone could be about being on the Moon, suggested that Schmitt take a moment to look at the Earth.

118:08:02 **Cernan:** "Oh, man. Hey, Jack, just stop. You owe yourself 30 seconds to look up over the South Massif and look at the Earth!"
118:08:07 **Schmitt:** "What? The Earth?"
118:08:09 **Cernan:** "Just look up there."
118:08:10 **Schmitt:** "Ah! You seen one Earth, you've seen them all."

Schmitt was kidding, but his concentration on his tasks was real. There would not be another scientist on the Moon for – who knew how long? He had to make every moment count.

Crew: Apollo 17

Mission Commander Eugene Cernan was an Apollo veteran, having flown on Apollo 10 (see p.39), the last set of tests in lunar orbit before the Apollo 11 landing. After John Young, he was the most seasoned lunar veteran ever to command an Apollo mission.

Jack Schmitt was born in New Mexico in 1935. He received a Bachelor of Science degree from Caltech in 1957 and studied at the University of Oslo in Norway between 1957 and 1958 on a Fulbright Scholarship. He received his Doctorate in Geology from Harvard University in 1964 and joined NASA in 1965.

The man who would mind the CM in orbit was Ronald Evans. Born in Kansas in 1933, he received a Bachelor of Science in Electrical Engineering from the University of Kansas in 1956 and a Master of Science in Aeronautical Engineering from the Naval Postgraduate School in 1964, where he became a close friend of Cernan. He was flying combat missions in Vietnam when notified of his acceptance by NASA in 1966.

The crew of Apollo 17. From the left: Jack Schmitt, Ron Evans and Gene Cernan (kneeling). Schmitt was the only scientist to fly to the Moon.

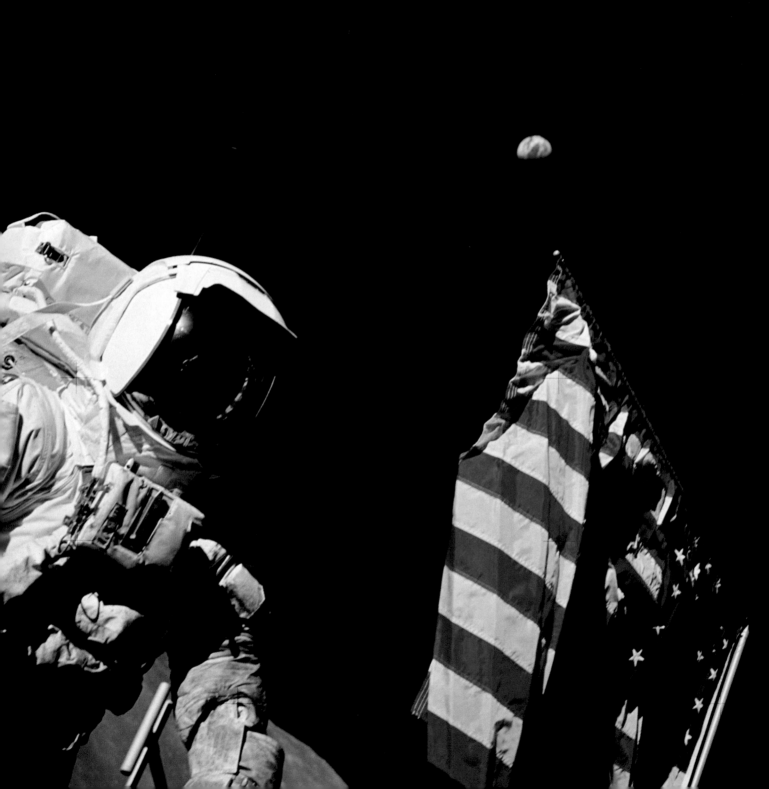

The landing area for this last exploration of the Moon had been the subject of great debate. There were just so many places the geologists wanted to explore and so many things left to do. With the cancellation of the three subsequent lunar missions (although the Apollo 18 hardware was eventually used with the Russian Soyuz 19 in Earth orbit in 1975) there were suddenly lots of targets to choose from.

Schmitt had hoped to go to a site on the far side of the Moon – Tsiolkovsky Crater. Besides being constantly exposed to a higher rate of bombardment from space than the near side, it offered the opportunity to sample some of the oldest lunar rock. Ultimately, a site noted after the flight of Apollo 15 was selected. It was the Taurus-Littrow Valley, a five-mile (eight-kilometer) wide canyon that promised to be the last salvation for theories of volcanism on the Moon. Or so they hoped. It also had two large mountains – the North and South Massifs – which promised to yield some old-crust samples. There was some concern over the narrowness of the landing area, but Cernan was not about to let something as minor as a threading-the-needle landing stop him from making this last mission one for the books.

With great skill, Cernan guided the LM *Challenger* into the valley. Within hours, the two men, as different as Armstrong and Aldrin had been, but both absolutely devoted to the challenge ahead, left the craft. There could have been no better choice than the seasoned astronaut Cernan, who became the last man on the Moon. As driven as any test pilot, he also had a sense of the poet about him.

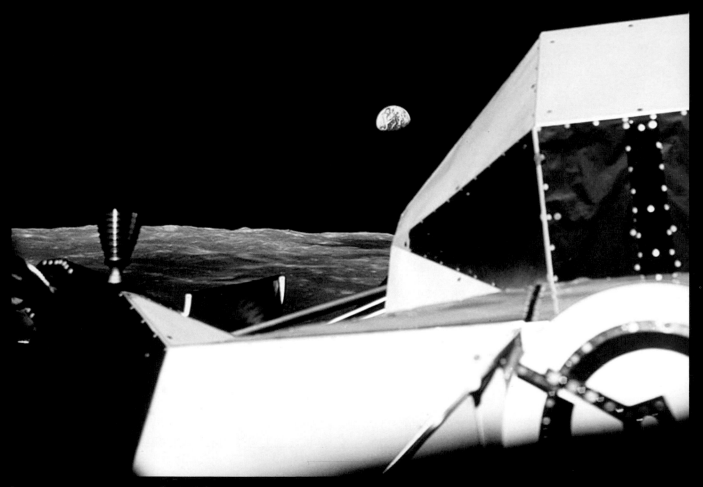

Earth as seen past the undocking Lunar Module *Challenger*.

The footpad of the last LM to land on the Moon — *Challenger*.

Jack Schmitt at the Rover. This is one of the only photos of an Apollo astronaut on the Moon's surface with his gold visor up. Schmitt's face can be seen inside the helmet.

The South Massif

With his vast experience as a test pilot and space traveller one might have expected Cernan to be a little more restrained on the Moon. But as he and Schmitt began their second EVA, which involved driving their Rover to the mountain known as the South Massif, he was clearly having a blast.

142:39:01 **Schmitt:** "That is a high mountain!"
142:39:03 **Cernan:** "Jimmeny Christmas!"

Cernan recalls feeling overwhelmed:
"The Massifs... rise maybe 8,000 feet straight up from the valley floor. What was really impressive about [them], however, was how massive they were... they overpowered you. Mt Rainier near Seattle gives you something of the same feeling... but Rainier is just one mountain and we were in a valley surrounded by jumbo mountains. On Earth, I've certainly never seen anything else like them."

142:44:27 **Schmitt:** "Look at Nansen!"

They were on top of a huge crater, Nansen, named after the Polar explorer.

142:44:31 **Schmitt:** "My goodness gracious."

They took some samples from the spectacular area, and moved on.

A few hours later, Schmitt made one of the most spectacular discoveries of the Apollo Program. They were exploring a location named Shorty Crater, when Schmitt's attention was riveted to a spot in the lunar soil near his feet.

145:26:25 **Schmitt:** "Wait a minute... "
145:26:26 **Cernan:** "What?"
145:26:27 **Schmitt:** "Where are the reflections? I've been fooled once."

Schmitt took a quick look around, checking for any false reflections emanating from the shiny Rover or their tools. He saw none. His eyes returned to the remarkable sight before him.

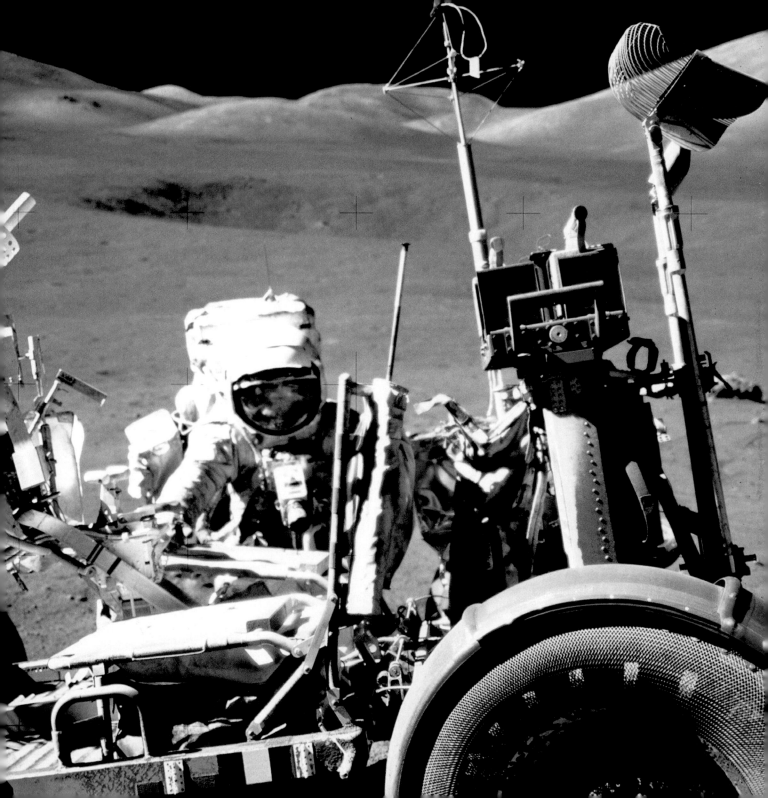

Orange soil on the Moon – the area was covered with glass beads, born of a volcanic "fire fountain."

145:26:27 **Schmitt:** "There is orange soil!"
145:26:32 **Cernan:** "Well, don't move it until I see it."
145:26:35 **Schmitt:** "It's all over! Orange!"
145:26:38 **Cernan:** "Don't move it until I see it!"

Cernan later recalled:
"Quite frankly, when Jack said he saw orange soil, I began to wonder if he hadn't been on the Moon too long, until I saw it myself."

145:26:40 **Schmitt:** "I stirred it up with my feet!"
145:26:42 **Cernan:** "Hey, it is! I can see it from here!"
145:26:44 **Schmitt:** "It's orange!"
145:26:46 **Cernan:** "Wait a minute, let me put my visor up."

A quick look confirmed the discovery.

145:26:46 **Cernan:** "It's still orange!"
145:26:49 **Schmitt:** "Sure it is! Crazy!"
145:26:53 **Cernan:** "Orange!"
145:26:54 **Schmitt:** "I've got to dig a trench, Houston."

The two quickly dug a trench and sank a core tube into the soil. They bagged all the samples they could in the time they had left.

145:27:00 **CapCom:** "Copy that. I guess we'd better work fast."
145:27:01 **Cernan:** "Hey, he's not going out of his wits. It really is."
145:27:07 **CapCom:** "Is it the same color as cheese?"

Schmitt took the remark in his stride, as he later recalled:
"I think [the CapCom] was thinking of green cheese, as in 'the Moon's made out of green cheese.' My guess is that he wasn't convinced that we weren't trying to pull their leg again. And for good reason, because every once in a while we'd do that."

145:27:15 **Schmitt:** "It's almost the same color as the LMP decal on my camera."

The jokes from the ground stopped, as the realization sank in.

145:27:21 **CapCom:** "Okay. Copy that."
145:27:23 **Cernan:** "That is orange, Jack! How can there be orange soil on the Moon? Jack, that is really orange. It's been oxidized..."
145:28:39 **Schmitt:** "It looks just like an oxidized desert soil, that's exactly right."

They both marveled at the sight. In an otherwise monochromatic landscape, here was a bright, vibrant color. Other missions had found some rocks of varying hue, but nothing like this.

145:29:15 **Schmitt:** "You know that orange is along a line, Geno, along the rim crest."
145:29:27 **Cernan:** "What? Circumferential?"
145:29:29 **Schmitt:** "Yeah. Man, if there ever was a... I'm not going to say it."

Schmitt bags a sample of Moon rock. The astronauts' stiff suits made working alone very difficult, but in the interest of saving time, they often did.

Schmitt took a moment here, to consider what he was about to say. It was sticking his neck out as a scientist to suggest that this might be a sign of volcanic gas, but...

145:29:29 **Schmitt:** "... if there ever was something that looked like [volcanic] alteration, this is it."

It transpired that the rocks were the result of a huge and violent geological event. Three-and-a-half billion years before, right where the two astronauts stood, a huge volcanic plume had thrust molten rock into the airless skies of the Moon. Some of the gas that erupted formed into orange glass beads as it cooled. The beads fell to the ground and were later covered by other gray, dusty deposits. It was pure luck that Schmitt scuffed the dirt here to discover the results of this tumultuous event. He later recalled:

"... normally, on Earth, that alteration is [an] alteration from oxidation. And so, when we saw orange, that was the immediate thought that everybody had. 'My God, there's a volcanic emanation here that's altered the soil.' Well, it turned out that that wasn't true. It was volcanic material, but it was volcanic glass that had been spewed out of some fire-fountain-like eruptions three-and-a-half billion years ago."

So it was not the kind of huge, cataclysmic and world-changing volcanic event that they had hoped to find. But science, and space, was like that. You had to accept what came your way.

Parked on an incline near Station 6. The boulder
was so large that it could be seen from orbit.

The Boulder

Near the end of their third EVA, the pressure was on.
The last moments of Apollo on the Moon were upon
them. As the Rover returned them to the South
Massif area of the Taurus-Littrow Valley, Schmitt
worried about time – there was so little of it, and so
much yet to explore.

They descended a slope to look at a boulder that
was so huge it had been spotted from orbit. The
slope was so severe that both men felt they might
tumble down the hill at any moment.

164:51:33 **Schmitt:** "You parked on a slope."
164:51:35 **Cernan:** "There's no level spot to park here."
164:51:40 **Schmitt:** "You want some help getting off?"
164:51:42 **Cernan:** "I've got to go uphill!"
164:51:44 **Schmitt:** "I just about ended up down at
the bottom of the hill."

Both astronauts were having some difficulty getting
off the Rover due to the steep gradient. Although

they chuckled as they struggled, their laughter hid the
concern both were feeling, which was magnified
tenfold at Mission Control.

164:52:14 **Schmitt:** "You want me to block the
wheels? You got the brake on, I hope."
164:52:20 **Cernan:** "You betcha! I don't know if I can
lean uphill enough! I can't. Holy Smoley! Boy, are we
on a slope!"
164:52:35 **Schmitt:** "You okay?"
164:52:36 **Cernan:** "Yeah. Let me get this thing set again."
164:52:38 **Schmitt:** "I don't think you can get a... "
164:52:40 **Cernan:** "Boy, are we on a slope!"

Cernan later recalled:
*"I had to get off [the Rover] uphill, and it was really
pretty hard to get off. Jack said he nearly rolled to
the bottom of the hill. It was almost like parking in
San Francisco."*

Schmitt trotted to the enormous boulder, which had a
large gash in the middle.

A long way back: the LM *Challenger* can be seen, tiny, at the center of this image. The explorations of Apollo 17 were the farthest any astronaut ventured from an LM.

Opposite: Cernan pulls a protective cover off the dedication plaque on the *Challenger*. He would, in minutes, be the last man on the Moon.

164:52:44 **Schmitt:** "Okay. I'm going to stay out from between the rocks. It's a beautiful east–west split rock. It's even got a north overhang that we can work with…"

Cernan was still distracted.

164:53:08 **Cernan:** "Oh, man, what a slope!"

A piece of the huge rock was a must-have sample. Schmitt's hands were so tired from the awkward spacesuits and the previous moonwalks that he asked Cernan to do the honors. The Commander took a few hefty swings at the massive boulder, and billions of years of history were shattered open.

Then, Cernan noticed some damage to the rock further up.

165:08:42 **Cernan:** "Looks like somebody's been chipping up there."
165:08:44 **Schmitt:** "Looks like there's been a geologist here before us!"

This rock, which had tumbled from the ridge above nearly a million years before, was formed of events from a massive impact nearly four billion years ago. It was huge, towering above both men. The sample they obtained painted a clear picture of the sequence of events leading up to the formations around them, and indeed of the formation of the entire Sea of Serenity, a 450-mile (725-kilometer) wide formation nearby.

The Last Goodbye

The astronauts returned to the LM and, after securing the rock samples, Schmitt headed up the ladder. He was bone tired, yet still reluctant to leave. Cernan was, for a brief moment, the very last man on the Moon. Although it was a moment of some gravity, few realized that 35 years later he would still hold that record. Cernan turned to the TV camera, now parked some distance away on the Rover, and spoke to the CapCom, although it was really a message to the world.

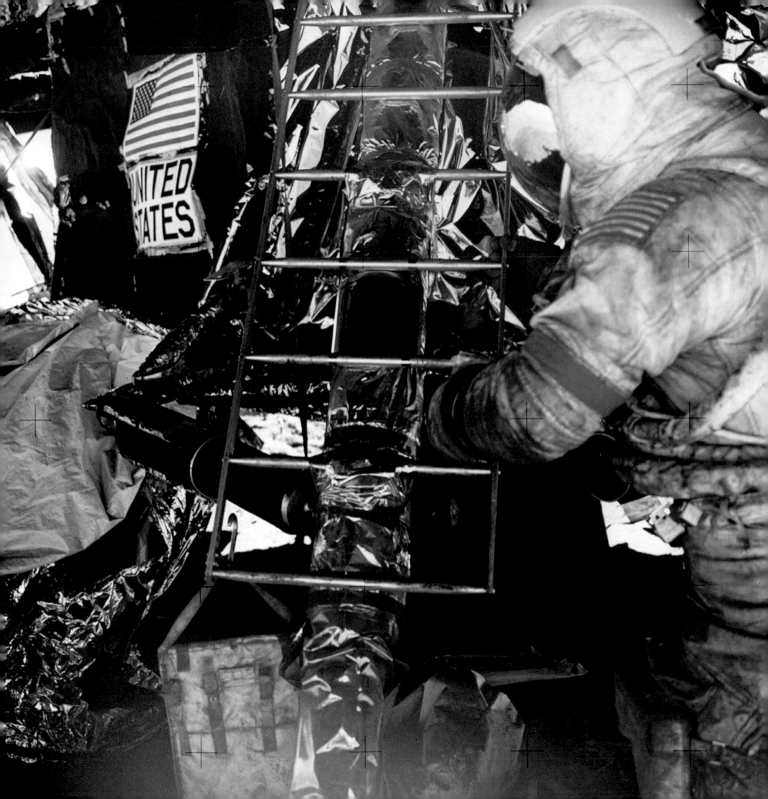

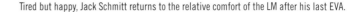
Tired but happy, Jack Schmitt returns to the relative comfort of the LM after his last EVA.

170:41:00 **Cernan:** "Bob, this is Gene, and I'm on the surface; and, as I take man's last step from the surface, back home for some time to come – but we believe not too long into the future – I'd like to just say what I believe history will record. That America's challenge of today has forged man's destiny of tomorrow. And, as we leave the Moon at Taurus-Littrow, we leave as we came and, God willing, as we shall return, with peace and hope for all mankind. Godspeed the crew of Apollo 17."

And with that, he went inside the LM and closed the hatch, joining Schmitt for their final hours on the Moon. Together, the test pilot and the geologist had pulled off a record mission.

There had been much debate in the NASA community about scientist-astronauts. Some felt that all astronauts should be test pilots, others felt that most should be scientists. The majority held opinions somewhere in-between. But Apollo 17 was the proof: there was no substitute for taking a scientist to the Moon. Jack Schmitt had proved his worth, and whether it be the shuttle or the space station, the future for these precisely qualified experts was assured. Scientists would, from now on, be included on every NASA mission. It was a fitting testament to the flights of Apollo.

The last Apollo capsule to voyage to the Moon, the *America*, splashes down within sight of the USS *Ticonderoga*.

01 THE FINAL WORD

Present Day: Earth

Al Bean of Apollo 12 was one of the most visually inclined of the astronauts to travel to the Moon. He is now a full-time artist, creating paintings that evoke that part of his extraordinary life. Captain Bean's new mission is to share the excitement and wonder of spaceflight with an earthbound audience. And he, as well as the other Apollo astronauts, share their thoughts on the future.

Each astronaut returned from the Moon a changed man; with something added to that particular set of characteristics that allowed him to go there in the first place. For each of them it was different. They were all overachievers and all strove for excellence. So it was natural to expect them to excel in some new venture when they returned to Earth. Many went into business and most were quite successful. Frank Borman took the helm of Eastern Airlines. Jim Irwin and Charlie Duke found themselves preaching the Gospels. Ed Mitchell founded an institute devoted to the study of

human consciousness. Michael Collins ran the Smithsonian National Air and Space Museum. And at least one found his calling in the world of art.

When Al Bean found himself on the Moon during the flight of Apollo 12, he had already come a long way from his early days in naval aviation. He had a calm determination and always set about doing the best job he could with whatever task he found himself assigned. This did not go unnoticed, and eventually landed him in the Lunar Module pilot's seat for the Apollo 12 mission. In 1973, he commanded the second Skylab flight, logging 59 days in orbit. But by 1981 a different calling came to him. For years he had studied painting in his spare time, and had applied his dogged determination to mastering oils.

That's How It Felt to Walk on the Moon: One of Bean's first paintings, and still a personal favorite. It captures his feelings about the Apollo Program in spectacular fashion.

He decided that, having been to places that few others had been and seen things few others had seen, it was time to bring his unique perspective to the world of artistic endeavor.

Changing Perspectives

In orbit around the Moon, Bean had made the decision that led to his new undertaking, He determined that from then on, he would live *exactly as he wanted to live*. Many years later, in the mid-1980s, he got serious about professional art. His artworks soon attracted attention, and today his paintings command substantial sums of money. But his reason for creating them remains unchanged: to bring the wonder of spaceflight to those who dream of it but cannot go. One of his earliest works, which is titled *That's How It Felt to Be on the Moon*, is still one of his favorites.

"I took one of the photos that Pete Conrad had taken of me, and said to myself, 'I'm going to paint that so that it has just the right feeling...'"

The result was an immediately popular rendition of a lone astronaut standing on the Moon, infused with the pastel colors of a painting by the French impressionist artist Claude Monet. It was not a typical interpretation of the harsh lunar environment.

"I painted it gold and yellow underneath the surface, to give it a glow, and it didn't look right. So I thought about it for perhaps a month, then looked at it again. What it needed was more of a rainbow. That mission was an exciting time, a rewarding time. But it's really hard to paint emotions. So I put those other colors in there, and it seemed to have more of the effect I was after. A more fanciful and ephemeral effect than reality. It turned out to make a lot of sense to other people, and it certainly did to me."

Bean's body of work is based on some of his first views of the Moon.

"When we landed, and first looked out the window, it was so very bright. It was just a white surface, like an incredibly bright day. It wasn't a black-and-gray world. It was almost too radiant to look at."

His impressions were not unlike Buzz Aldrin's first observation: "magnificent desolation."

"You tend to think of desolation as some sort of a uniform thing.... I often think of it as meaning nothing. But there were always craters and rock and shadows. And it was different from day to day. Depending on the angle of the Sun in the lunar sky, the landscape would appear either gray or tan. I remember on our second moonwalk, I noticed that one of the rocks I had inspected the day before was a different color. I got excited, but then looked around and noticed that all the rocks were a different color. It was the light at work."

One of the things Bean found most remarkable was of earthly manufacture – the Lunar Module itself. After the trek to Surveyor 3, the LM presented a remarkable sight.

"Seeing the Surveyor was a bit like seeing an old rusted car in the woods. It was an artefact. But the LM was so much bigger than us. When you stepped back from it, it looked like you just landed your house on the Moon. A house wrapped in gold foil. It seemed out of place."

Then, reflectively, he added:
"It's comforting to have something that big to return to when you're exploring the Moon."

Another one of Bean's classic paintings. *Heavenly Reflections* shows the strong feeling of camaraderie among US astronauts. Again, the use of color and angle inspires feelings of strength, friendship and wonder.

Of all the Apollo crews that landed on the Moon, perhaps Bean and Pete Conrad, his commander, had the best time during their short stay.

"I was lucky to be with Pete. I was always glad to be on his crew. We were good friends forever. Who knows why people connect like that. All the astronauts were competent, and any one of us could have done the job. But I was just glad to be around Pete."

A Philosophy

The road to the Moon had not been an easy one for Bean. He had always worked harder than many of the other guys. He often felt that some of them were more naturally gifted as pilots and military men, so he vowed to make up for it by dint of hard work. This, he feels, is what it takes to go to the Moon.

"There are a lot of things that we have to do to have a good life. I always felt that you must love what you are doing, and have people you love, and people who love you. If one of those things is missing, then it gets tough. It's good to have a road map for life, because people need them to follow their dreams. Just wanting is not enough. I think I was lucky, too. I was the right age, at the right place, at the right time. Now it's not quite the same. Even the very best test pilot can't go to the Moon today – nobody can. I look back and it seems like it was a very special time."

Bean thinks about that a lot. There is no regret or sadness, just a feeling of great fortune.

"I feel lucky every day to have done what I have done, and to be doing what I am doing. I feel fortunate two or three times a day."

Then, with a wink he adds:
"But then, you have to be a person who finds luck in whatever you do."

The Road Ahead

Bean does not see an easy time ahead:

"We were racing the Russians; trying to build a bigger rocket and get there first. But that was not the real motivating force for those of us at NASA."

Bean feels that NASA and the astronauts had a dream that the powers in Washington and the nation at large may no longer share.

"We had a dream, to go to the Moon, to land there and return to Earth. That's what surrounded me for so many years. And I think that's why we are not going there, or to Mars, now. There is not that motivating force. There's always that notion of trying to solve the problems close at hand, on Earth, first. You can have the vision, but without the financing, you are not going to go. I don't think we will do it again until there is a compelling reason. Or perhaps it will be a breakthrough in propulsion; a faster and less expensive way to get there. But I don't think I'll see it in my lifetime."

The Apollo Program lasted an all-too-brief 11 years. When it was canceled, nobody at NASA, even the pessimists, thought that 35 years would pass without another manned mission leaving Earth orbit. But that is exactly what has occurred. When you talk to the other Apollo astronauts, their opinions about the future vary widely. But there are common threads:

We Must Return: The United States had just started a successful program of exploration of the Moon. It should

continue. And most feel that there is a place for a lunar base as a staging point for a Mars mission, although some feel that Earth orbit would work just as well.

We Must Go to Mars: Mars is the next logical target. The only other place nearby is the asteroid belt. But while those bodies are interesting and possibly resource-rich, they do not have the grandeur of Mars. And grandeur feeds the human soul.

We Need Improved Technologies: Only through improved propulsion and engineering will spaceflight become affordable, and perhaps even profitable. This requires government intervention.

And...

If We Don't Go, Someone Else Will: Just who might be the United States' primary competitor varies according to whom you speak. Some feel that it will be the Chinese, others cite the European Union. But whoever it is, the United States will have lost an opportunity.

The Apollo Program involved nearly half a million people and over 20,000 companies working together in a harmony unmatched in the twentieth century. It was the finest creation of a peacetime civilization. And it performed with unmatched success.

Perhaps Buzz Aldrin sums it up best, when he paraphrases his own historic words:

"We need to move from Magnificent Desolation to Magnificent Inspiration."

The future awaits the brave.

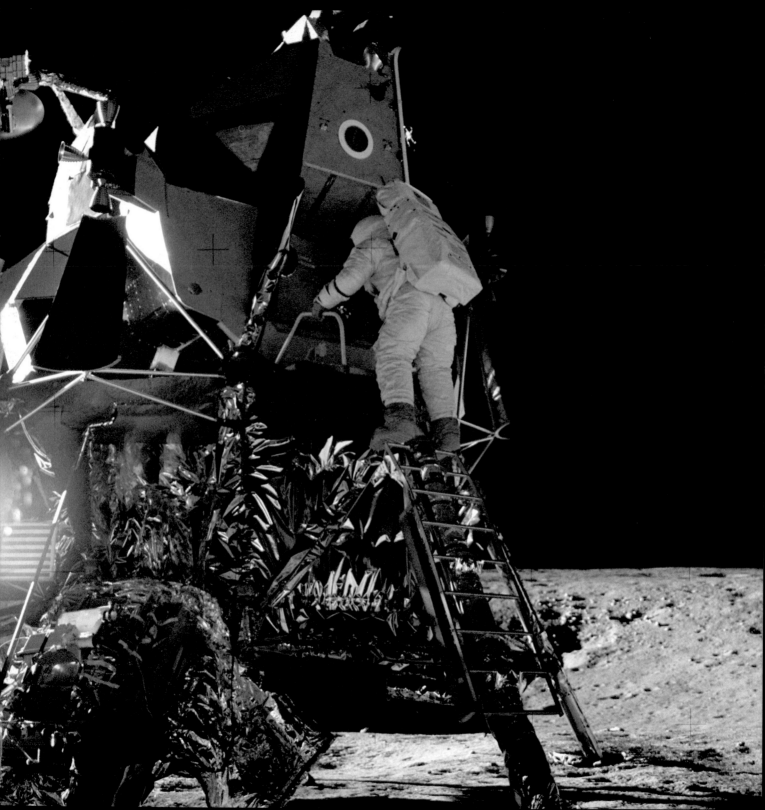

APPENDIX

Mission data for US manned spaceflight programs: Project Mercury to Apollo.

Note on orbits: Where two numbers are specified to measure altitude, the larger number is the orbital *apogee*, which is the farthest point the spacecraft reached from the Earth or Moon. The smaller number is the *perigee*, or closest point. Two numbers are used because the orbits were elliptical.

Project Mercury

NB: MR=Mercury capsule on a Redstone booster rocket; MA=Mercury Capsule on Atlas rocket.

MR-3 *Freedom 7*
Crew: Alan B. Shepard, Jr
Objective: The main scientific objective of Project Mercury was to determine human capabilities in the environment of space and while traveling there and back.
Launch: 5 May, 1961.
Altitude: 116.5 miles (187.5 kilometers).
Duration: 15 min, 28 sec.
Distance: 303 miles (487.6 kilometers).
Velocity: 5,134 miles per hour (8,262 kilometers per hour).
Splashdown: Atlantic Ocean.

MR-4 *Liberty Bell 7*
Crew: Virgil I "Gus" Grissom.
Objective: This was the fourth mission in the Mercury-Redstone series of flight tests and the second American manned sub-orbital spaceflight.
Launch: 21 July, 1961.
Altitude: 118.3 miles (190.4 kilometers).
Duration: 15 min, 37 sec.
Distance: 302 miles (486 kilometers).
Velocity: 5,134 miles per hour (8,262 kilometers per hour).
Splashdown: Atlantic Ocean, 302 miles (486 kilometers) east of launch site.

MA-6 *Friendship 7*
Crew: John H. Glenn, Jr. First American in Earth orbit.
Objective: To place an astronaut into orbit, observe his reactions and safely return him to Earth to a point where he could be readily found.
Launch: 20 February, 1962.
Altitude: 162.2 by 100 miles (261 by 160 kilometers).

Orbits: Three, each lasting 88 min 29 sec.
Duration: Four hours, 55 min, 23 sec.
Distance: 75,679 miles (121,790 kilometers).
Velocity: 17,544 miles per hour (28,233 kilometers per hour).
Splashdown: 800 miles (1,287.5 kilometers) southeast of Bermuda.

MA-7 *Aurora 7*
Crew: M. Scott Carpenter.
Objective: To corroborate the previous mission.
Launch: 24 May, 1962.
Altitude: 166.8 by 99.9 miles (268.4 by 160 kilometers).
Orbits: Three of 88 min 32 sec.
Duration: Four hours, 56 min, five sec.
Distance: 76,021 miles (122,341 kilometers).
Velocity: 17,549 miles per hour (28,242 kilometers).
Splashdown: Atlantic Ocean. Spacecraft overshot intended target area by 250 miles (400 kilometers).

MA-8 *Sigma 7*
Crew: Walter M. Schirra, Jr.
Objective: To place an astronaut in orbit for a greater length of time.
Launch: 3 October, 1962.
Altitude: 175.8 by 100 miles (283 x 160 kilometers).
Orbits: Six of 88 min 55 sec.
Duration: Nine hours, 13 min, 11 sec.
Distance: 143,983 miles (231,712 kilometers).
Velocity: 17,558 miles per hour (28,256 kilometers per hour).
Splashdown: Pacific Ocean.

MA-9 *Faith 7*
Crew: L. Gordon Cooper.
Objective: To accomplish a one-day mission in orbit.
Launch: 15 May, 1963.
Altitude: 165.9 by 100.3 miles. 267 by 161.4 kilometers).
Orbits: 22.5 of 88 min 45 sec.
Duration: One day, 10 hours, 19 min, 49 sec.
Distance: 546,167 miles (878,947 kilometers).
Velocity: 17,547 miles per hour (28,238 kilometers per hour).
Splashdown: After 22 orbits, virtually all spacecraft systems had failed. Cooper fired the retrorockets manually and the spacecraft re-entered the atmosphere, landing safely in the

Pacific Ocean 34 hours, 19 min and 49 sec after liftoff.

Project Gemini

NB: all missions used a Titan-II rocket to launch the capsules.

Gemini 3

Crew: Virgil I. Grissom, John W. Young.
Objective: To demonstrate manned orbital flight and evaluate two-man capsule design.
Launch: 23 March, 1965.
Altitude: 139 miles (224 kilometers).
Orbits: Three.
Duration: Four hours, 52 min, 31 sec.
Splashdown: Atlantic Ocean. Miss landing zone by 60 miles (111.1 kilometers).

Gemini IV

Crew: James A. McDivitt, Edward H. White II.
Mission Objective: To evaluate the effects of prolonged spaceflight. First NASA EVA.
Launch: 3 June, 1965.
Altitude: 184 miles (296.1 kilometers).
Orbits: 62.
Duration: Four days, one hour, 56 min, 12 sec.
Splashdown: 7 June, 1965. Atlantic Ocean.

Gemini V

Crew: C. Gordon Cooper, Charles Conrad.
Objective: To evaluate the rendezvous guidance and navigation system and demonstrate eight-day capability of spacecraft and crew.
Launch: 21 August, 1965.
Altitude: 217.3 miles (349.8 kilometers).
Orbits: 120.
Duration: Seven days, 22 hours, 55 min, 14 sec.
Splashown: August 29, 1965. Atlantic Ocean.

Gemini VII

Crew: Frank Borman, James A. Lovell.
Objective: Primary object was to conduct a 14-day mission and evaluate the effects on the crew. The craft was also to provide a docking target for Gemini VI.
Launch: 4 December, 1965.
Altitude: 203 miles (327 kilometers).
Orbits: 206.

Duration: 13 days, 18 hours, 35 min, one sec.
Splashdown: 18 December, 1965. Atlantic Ocean.

Gemini VI

Crew: Walter M. Schirra, Thomas P. Stafford.
Objective: Scheduled to be launched before Gemini VII and rendezvous with another unmanned target. Unmanned Agena rocket intended as rendezvous target exploded shortly after liftoff, so Gemini VI was postponed and rescheduled to follow Gemini VII into space and rendezvous with it.
Launch: 15 December, 1965.
Altitude: 168 miles (311.3 kilometers).
Orbits: 16.
Duration: one day, one hour, 51 min, 24 sec.
Splashdown: 16 December, 1965. Atlantic Ocean.

Gemini VIII

Crew: Neil A. Armstrong, David R. Scott.
Objective: To rendezvous and dock with Gemini Agena target vehicle.
Launch: 16 March, 1966. 11:41:02.389.
Altitude: 185.6 miles (298.7 kilometers).
Orbits: Seven.
Duration: 10 hours, 41 min, 26 sec.
Splashdown: 17 March, 1966. Pacific Ocean.

Gemini IX-A

Crew: Thomas P. Stafford, Eugene A. Cernan.
Objective: Primary objective was to perform rendezvous and docking and conduct an EVA.
Launch: 3 June, 1966.
Altitude: 193.5 miles (311.5 kilometers).
Orbits: 45.
Duration: Three days, 20 min, 50 sec.
Splashdown: 6 June, 1966. Atlantic Ocean.

Gemini X

Crew: John W. Young, Michael Collins.
Objective: The primary objective was to rendezvous and dock with a Gemini Agena target vehicle.
Launch: 18 July, 1966.
Altitude: 468 miles (753.3 kilometers).
Orbits: 43.

Duration: Two days 22 hours 46 min 39 sec.
Splashdown: 21 July, 1966. Pacific Ocean.

Gemini XI
Crew: Charles Conrad J., Richard F. Gordon Jr.
Objective: The primary objective was to rendezvous and dock with Gemini Agena target vehicle.
Launch: 12 September, 1966.
Altitude: 850.6 miles (1,368.9 kilometers).
Orbits: 44.
Duration: Two days 23 hours 17 min, eight sec.
Splashdown: 15 September, 1966. Pacific Ocean.

Gemini XII
Crew: James A. Lovell Jr., Edwin E. Aldrin.
Objective: The primary objectives were rendezvous and docking and EVA evaluation.
Launch: 11 November, 1966.
Altitude: 187 miles (301.3 kilometers).
Orbits: 59.
Duration: Three days, 22 hours, 34 min, 31 sec.
Splashdown: 15 November, 1966. Pacific Ocean.

The Apollo Program

Apollo 1 (204) Saturn 1B rocket
Crew: Virgil "Gus" Grissom, Edward White II, Roger Chaffee.
Objective: Intended as first Apollo flight crew and flight of early (Block 1) Command Module. Caught fire and burned on the pad. All three astronauts perished in the fire. Flight number posthumously changed from Apollo 204 to Apollo 1.

Apollo 7 Saturn 1B AS-205
Crew: Walter M. Schirra, Jr, Donn F. Eisele, R. Walter Cunningham.
Objectives: This was the first piloted Command Module (CM) mission and first three-man American space crew. The primary objectives for the mission were to demonstrate CM/crew performance; demonstrate crew/space vehicle/mission support facilities' performance during a piloted Command/Service Module (CSM) mission and demonstrate CSM rendezvous capability. The mission provided the first live TV downlink from space.
Launch: 11 October, 1968.
Altitude: 140 by 183 miles (225.3 by 294.5 kilometers).
Orbits: 163.
Duration: 10 days, 20 hours.
Splashdown: 21 October, 1968. Atlantic Ocean. The CSM propulsion system, which had to fire the craft into and out of Moon orbit, worked perfectly during eight burns lasting from half a second to 67.6 sec.

Apollo 8 Saturn V AS-503
Crew: Frank Borman, Commander, James A. Lovell, Jr, William A. Anders.
Objectives: The mission aimed to demonstrate the crew/space vehicle/mission support facilities during a manned Saturn CSM mission. It also aimed to demonstrate trans-lunar injection, CSM navigation, communications and midcourse corrections and return high-resolution photographs of proposed Apollo landing sites and locations of scientific interest.
Launch: 21 December, 1968.
Altitude: 118 by 112 miles (190 by 180 kilometers).
Duration: Six days, 3 hours, 42 sec.
Distance: Lunar orbit; 243,000 miles (391,230 kilometers).
Splashdown: 27 December, 1968. Pacific Ocean.

Apollo 9 Saturn V AS-504
Crew: James A. McDivitt, David R. Scott, Russell L. Schweickart.
Objectives: To demonstrate crew/space vehicle/mission support facilities during manned Saturn CSM/Lunar Module (LM) mission, the LM/crew performance and the selected lunar orbit rendezvous mission activities including transposition, docking withdrawal, inter-vehicular crew transfer, and Lunar Module active rendezvous and docking.
Launch: 3 March, 1969.
Altitude: 119 miles (192 kilometers). Earth orbit.
Duration: 10 days, one hour.
Distance: 123 x 127 miles (198 x 204 kilometers).
Splashdown: 13 March, 1969. Atlantic Ocean.

Apollo 10 Saturn V AS-505
Crew: Eugene A. Cernan, John W. Young, Thomas P. Stafford
Objectives: To demonstrate the performance of LM and CSM in the lunar gravitation field and evaluate the lunar navigation of the CSM and LM docked and undocked.
Launch: 18 May, 1969.
Altitude: 118 by 114 miles (190 kilometers x 184 kilometers). Earth orbit.
Duration: Eight days, three min, 23 sec.
Distance: Lunar orbit; 243,000 miles (391,230 kilometers).
Splashdown: 26 May, 1969. Pacific Ocean.

Apollo 11 Saturn V AS-506
Crew: Neil A. Armstrong, Edwin E. Aldrin, Jr., Michael Collins.
Objective: To perform manned lunar landing and return safely to Earth.
Launch: 16 July, 1969.
Altitude: 115.5 by 113.7 miles (186 by 183 kilometers). Earth orbit.

Duration: Eight days, three hours, 18 min, 35 sec.
Distance: Lunar orbit and landing; 243,000 miles (391,230 kilometers).
Lunar landing location: Sea of Tranquility.
Lunar Co-ordinates: 71° north, 23.63° east.
Splashdown: 24 July, 1969. Pacific Ocean.

Apollo 12 Saturn V AS-507
Crew: Charles Conrad, Jr., Richard F. Gordon, Jr., Alan L. Bean.
Objectives: The second manned mission to the Moon. Goals included a more precise landing and collection of hardware from Surveyor III, which landed on the Moon 3 years earlier.
Launch: 14 November, 1969.
Duration: 10 days, four hours, 36 min.
Distance: Lunar orbit; 243,000 miles (391,230 kilometers).
Lunar landing location: Ocean of Storms.
Lunar co-ordinates: 3.04° south, 23.42° west.
Splashdown: 24 November, 1969. Pacific Ocean.

Apollo 13 Saturn-V AS-508
Crew: James A. Lovell, Jr, John L. Swigert, J., Fred W. Haise, Jr.
Objective: Apollo 13 was supposed to land in the Fra Mauro area. An explosion on board crippled the craft and forced it to circle the Moon without landing. The Fra Mauro site was reassigned to Apollo 14.
Launch: Saturday, 11 April, 1970.
Duration: Five days, 22 hours, 54 min.
Distance: Lunar orbital pass; 243,000 miles (391,230 kilometers).
Splashdown: 17 April, 1970. Pacific Ocean.

Apollo 14 Saturn-V AS-509
Crew: Alan B. Shepard, Jr, Stuart A. Roosa, Edgar D. Mitchell.
Objectives: The first flight after Apollo 13 took the aborted mission's landing site, Fra Mauro. The mission proved to be a successful return to the Apollo Program's goals.
Launch: 31 January, 1971.
Duration: Nine days.
Distance: Lunar orbit; 243,000 miles (391,230 kilometers).
Lunar Location: Fra Mauro.
Lunar Co-ordinates: 3.65° south, 17.48° west.
Splashdown: February 9, 1971. Pacific Ocean.

Apollo 15 Saturn V AS-510
Crew: David R. Scott, James B. Irwin, Alfred M. Worden.
Objectives: The mission accomplished three EVAs, totaling 10 hours, 36 minutes. The scientific payload landed on the Moon doubled. Improved spacesuits gave increased mobility and allowed a longer stay on the lunar surface (66.9 hours). The Lunar Roving Vehicle (LRV traversed total 17 miles (27 kilometers). The CSM was in lunar orbit for 145 hours, and

underwent 74 orbits. 169 pounds (76 kilograms) of lunar material was gathered.
Launch: 26 July, 1971.
Duration: 12 days, 17 hours, 12 min.
Distance: Lunar orbit; 243,000 miles (391,230 kilometers).
Lunar Location: Hadley-Apennine.
Lunar Co-ordinates: 26.08° north, 3.66° east.
Landing: 7 August, 1971. Pacific Ocean.

Apollo 16 Saturn V AS-511
Crew: John W. Young, Thomas K. Mattingly II, Charles M. Duke, Jr.
Objectives: This mission was to be the first to study the lunar highlands. An ultraviolet camera/spectrograph was used for first time on the Moon and the LRV was used for the second time. The time spent on the lunar surface was 71 hours, while the time spent in lunar orbit was 126 hours, with 64 orbits. 213 pounds (95.8 kilograms) of lunar samples were collected.
Launch: 16 April, 1972
Duration: 11 days, one hour, 51 min, sec.
Distance: Lunar orbit; 243,000 miles (391,230 kilometers).
Lunar Location: Descartes Highlands.
Lunar Co-ordinates: 8.97° south, 15.51° east.
Landing: 27 April, 1972. Pacific Ocean.

Apollo 17 Saturn-V AS-512
Crew: Eugene A. Cernan, Ronald E. Evans, Harrison H. Schmitt.
Objectives: The lunar landing site was the Taurus-Littrow highlands and valley area. This site was picked for Apollo 17 because it was a location where rocks that were both older and younger than those previously returned from other Apollo missions and from the Luna 16 and 20 unmanned missions might be found. The mission was the final in a series of three J-type missions planned for the Apollo Program. These J-type missions can be distinguished from previous G and H-series missions by extended hardware capability, larger scientific payload capacity and the use of the battery-powered Lunar Roving Vehicle (LRV). The scientific objectives of the Apollo 17 mission included geological surveying and sampling of materials and surface features in a pre-selected area of the Taurus-Littrow region, deploying and activating surface experiments and conducting inflight experiments and photographic tasks during lunar orbit and trans-Earth coast (TEC). The lunar surface time was 75 hours.
Launch: 7 December, 1972.
Duration: 12 days, 13 hours, 52 min.
Distance: Lunar orbit; 243,000 miles (391,230 kilometers).
Lunar Location: Taurus-Littrow Valley.
Lunar Co-ordinates: 20.16° north. 30.77° east.
Splashdown: 19 December, 1972. Pacific Ocean near American Samoa.

GLOSSARY

A

Abort System: The Abort System included the equipment required to remove the Apollo CM from the Saturn V rocket if a launch went wrong. It was also known as the Escape Tower.

Acquisition of Signal (AOS): The time at which a signal from a spacecraft is received back on Earth.

Anorthosite: An intrusive igneous rock that primarily consists of calcium-rich plagioclase feldspar. Anorthosite was the much-sought primordial mineral of the lunar missions.

Apogee: The point in a satellite's orbit when it is farthest from Earth.

Apollo Lunar Scientific Experiment Package (ALSEP): Equipment that was similar to the EASEP (see below) but with more robust instrumentation and a nuclear power supply. See also RTG.

Attitude Control System (ACS): The spacecraft subsystem that is responsible for pointing the spacecraft in the desired direction and tracking it is pointed at all times.

B

Basalt: An extrusive igneous rock that is low in silica, dark in color and rich in iron and magnesium.

C

CapCom: The Capsule Communicator, usually another astronaut. During a mission, the CapCom is generally the person who speaks to the astronauts.

Circumlunar: A highly elliptical trajectory that follows a path round the Moon and returns to Earth.

Command Module (CM): The Apollo "capsule", which housed the astronauts. It was built by North American Rockwell.

Command and Service Module (CSM): The combined CM and Service Module (see below).

D

Display-Keyboard (DSKY): The Apollo guidance and navigation computer.

Downlink: A signal received from a spacecraft.

E

Early Apollo Scientific Experiment Package (EASEP): This contained various devices to send measurements back to Earth after the Apollo 11 mission had left the Moon.

Ejecta: Material thrown out by a volcano or cosmic impact. Ejecta includes pyroclastic material and, from some volcanoes, lava bombs. The term also includes material from the impacted area.

F

Fire fountain: A plume of incandescent molten lava sprayed vertically into the air from a volcanic vent or fissure.

G

Gnomon: The part of a sundial, usually a rod or fin pointed at the celestial pole, which casts the shadow. During the Apollo Program, a gnomon was used as a calibrating device for taking photos on the lunar surface.

H

Hydrazine: An early propellant that was soon replaced by unsymmetrical dimethylhydrazine (UDMH). It was used as a hypergolic fuel for rockets and maneuvering thrusters.

Hypergolic: Fuels that combust spontaneously upon contact, needing no ignition source. Used for the SM engine, the LM engines, and some RCS systems.

I

Igneous rock: Rock that has crystallized from a molten state.

Impact crater: A feature that has been created by the impact of cosmic bodies, for example meteorites, asteroids or comets, on a planet's surface.

J

Johnson Space Center (JSC): Located in Houston, Texas, JSC was also known as Mission Control.

K
Kennedy Space Center (KSC): This was where all Apollo launches originated.

L
Liquid oxygen: Used as an oxidizer for rockets and fuel cells.

Loss of Signal (LOS): When the reception of radio signals from a spacecraft is lost due to travel behind the Moon, ionization during re-entry or another anomaly.

Low Earth Orbit (LEO): Orbits that have apogees and perigees below about 1,800 miles (3,000 kilometers).

Lunar Module (LM): This was built by Grumman Aerospace and originally called the Lunar Excursion module (LEM). It was later shortened to LM.

Lunar Roving Vehicle (LRV): Commonly referred to as "the Rover," the LRV was built by Boeing and was used on Apollo 15, 16 and 17.

M
Mare, Maria: Large expanses of dark basalt on the Moon containing few craters. Nearly all of the mare are on the Moon's near side and most fill impact basins.

N
NASA: National Aeronautics and Space Administration.

P
Payload: The load carried by a spacecraft that includes passengers and cargo.

Pyroclastic: A material formed by a volcanic explosion or by aerial expulsion from a volcanic vent.

R
Rayed features: A term for the "splatter" pattern of ejecta (see above) after a crater impact. It is the bright material surrounding large fresh craters.

Reaction Control System (RCS): The "jets" that steered the Apollo spacecraft in space.

Regolith: The layer of loose rock resting on bedrock, constituting the surface of most land. On the Moon it refers to the surface soil and material.

Rift: A narrow fissure or other opening in a rock, caused by cracking or splitting. This fissure may allow magma to erupt on the surface as a lava flow.

Rilles: Long, narrow, trench-like valleys on the surface of the Moon, which may be straight or winding and that are likely to have been produced by the movement of lava. They have been compared to lava tubes or channels on Earth.

RP1: Kerosene used as rocket fuel. Used in the first stage (S1C) of the Saturn V rocket.

Radioisotope Thermoelectric Generator (RTG): These sub-critical nuclear power sources are ideal for planetary missions because of their long life and steady power in sunlight or darkness.

S
Service Module (SM): The cylindrical part of the Apollo spacecraft behind the CM (see above), which housed the propulsion and life support systems. The SM was by North American Rockwell.

T
Trans-Earth injection (TEI): Heading from the Moon to the Earth.

Trans-lunar injection (TLI): Heading from the Earth to the Moon.

U
Uplink: A signal sent to a spacecraft.

UTC: The worldwide scientific standard of timekeeping, or coordinated Universal Time. Also known as Greenwich Mean Time (GMT).

Z
Zenith: The point on the celestial sphere directly above the observer.

WEBSITES

http://www.hq.nasa.gov/alsj/: The home page for the Apollo Lunar Surface Journal by Eric Jones. The most comprehensive reference to the surface time of the Apollo missions.

http://www.astronautix.com/lvs/saturnv.htm: Home of Astronautix, the premiere Apollo (and space) encyclopedia.

http://www.space.com/: Your one-stop source for space news and history.

http://www.apollosaturn.com/: Great overall reference site. Details of all launches, hardware, and more.

http://spaceflight1.nasa.gov/history/apollo/: NASA's history of the Apollo Program, as well as Mercury, Gemini, Shuttle and Space Station missions. Highly interactive and searchable.

http://www.nasm.si.edu/collections/imagery/apollo/apollo.htm: The Smithsonian Institution's highly detailed reference of the Apollo Program.

http://www.apolloarchive.com/: While a bit less polished than the others, this site offers a veritable wealth of information, imagery, timelines and more. Most anything Apollo can be found here.

http://www.apolloarchive.com/apollo_retrospective.html: "Contact Light" is the title of this personal website. An excellent overview of lunar exploration from the personal point of view of the author.

http://nssdc.gsfc.nasa.gov/planetary/lunar/apollo.html: Another NASA-related site, this one includes a lot of specific data you might not find elsewhere.

http://science.ksc.nasa.gov/history/apollo/apollo.html: The Kennedy Space Center's Apollo site. Good links and references, as well as detailed mission histories.

http://www.hq.nasa.gov/office/pao/History/diagrams/apollo.html: Technical and engineering drawings of most Apollo flight hardware.

http://www.hq.nasa.gov/office/pao/History/diagrams/apollo.html: Johnson Space Center's extensive holdings of multimedia can be found here.

http://www.novaspace.com/: Home of Novagraphics, purveyors of the highest quality space photographs and artwork.

http://eaglelander3d.com/: A free lunar-landing simulator. Absolutely accurate in detail, right down to the readings on the navigation computer.

http://www.pbs.org/wgbh/nova/tothemoon/: A PBS site, includes a wealth of detail and activity.

http://smithsonianeducation.org/students/index.html: The Smithsonian's science site for kids, includes an Apollo 11 retrospective.

http://www.jpl.nasa.gov/index.cfm: Jet Propulsion Laboratory's site for all things unmanned. A great companion to manned spaceflight history.

http://www.sln.org/pieces/schutte/LMintro2.html: Specific information about the work of Grumman Aerospace on the Apollo Lunar Module.

http://www.boeing.com/history/bna/histspace.html: The history of North American Rockwell's involvement in Project Apollo.

http://www.spacecamp.com/: Home of Spacecamp, NASA's own spaceflight experience.

PICTURE CREDITS

Alan Bean: Alan Bean: 176, 179
Northrup Grumman: 104
National Aeronautics and Space Administration (NASA):
5, 10, 12, 14, 15, 16, 19, 20-21, 23, 24, 26-27, 28-29, 30-31, 32, 34-35, 36, 37, 38-39, 40, 41, 42, 44-45, 47, 50-51, 53, 54-55, 57, 58, 59, 61, 62, 64, 65, 66,68, 69-70, 71, 72, 75-75, 77, 78-79, 80, 81, 83, 84, 86, 87, 88-89, 90-91, 92-93, 94, 96-97, 98, 99, 100; /:102, 103, 105, 106, 108-109, 111, 113, 114, 115, 116, 118, 119, 121, 112, 124-125, 126-127, 129,130, 132-133, 134, 136, 137, 138, 140-141, 142, 144, 146, 149, 150-151, 152-153, 153 (inset), 155, 156-157, 158, 160-161, 163, 164, 166-167, 168, 170, 171, 172-173, 174;, 181.

United States Geological Survey (USGS), Kip Teague: 7

INDEX

Page numbers in bold refer to illustrations

ACKNOWLEDGMENTS

There are always many people to be grateful to in a project such as this, especially when one is dealing with facts on such a broad scale. First and foremost, I would like to express my profound thanks to Eric Jones, author, editor and curator of the spectacular Apollo Lunar Surface Journal, available on the Internet via various NASA websites. Of course, many other NASA sites were referenced for this book, and can be found in on page 188. All their curators have gone well beyond the call of duty in preparing their excellent compendiums of the Apollo era.

Kip Teague has kept the images of Apollo alive for a new generation in his incredible Apollo Image Gallery, also available on the Web. The time he has spent making this gift available to us is incalculable.

Andrew Chaikin, author of the wonderful book *A Man on the Moon* (which was later adapted by Tom Hanks for the mini-series *From the Earth to the Moon*) was supportive and helpful to a fault. His advice and kind words were an inspiration.

Piers Murray Hill and Amie McKee of Carlton Books were unfailingly helpful and positive. They are both a joy to collaborate with.

Marc Honorof of First Person Publishing and Rob Kirk and Rob Lihani of Digital Ranch are to be thanked for getting this project off the ground long after it had been a dream of mine. Don Cambou, creator of some of the best non-fiction television made, provided important support for my previous Apollo venture, *In Their Own Words: The Space Race*, and in doing so provided the impetus for this project.

Apollo moonwalker Alan Bean was a thrill to work with and has my gratitude. Charlie Duke, Gene Kranz and Gene Cernan were also gracious and generous with their time.

Kudos is also due to Peter Orton and Don Roberts, both Stanford folk, who helped to sharpen my writing skills, and Ron Monsen for creating the inspiring Eagle Lander 3D simulator. Ron Weber of SpaceShots.com gave advice on imaging that was appreciated.

Finally, Ken Kramer is due thanks for his unfailing support and advice. Jerry Marble was always there to lend an ear when it was most needed, as he has been for over 30 years. And most indispensable has been Gloria Lum, my lovely wife and life partner, who has weathered a dozen such projects and is always kind, loving and supportive when it is needed the most.